Infrastructure Development and its Environmental Impact

About the Author

Dr. Prabha Shastri Ranade, a Master's degree in Geography and Ph.D in Social Sciences from Nagpur University. She has over 25 years teaching and research experience. Presently, she is Consulting Editor at ICFAI Business School Research Centre, Ahmedabad. Starting her career as Research Officer at National Atlas and Thematic Mapping Organization, Kolkata, she moved to teaching of Geography at Nagpur University. Later, she was awarded UGC's Research Associateship for post-doctoral research projects, under which she worked at Devi Ahilya Vishwa Vidyalaya, Indore and Jawaharlal Nehru University (JNU), New Delhi for about 15 years. Dr. Ranade has contributed numerous articles in academic journals, conference proceedings, edited books, newspapers and the Encyclopdia of World Geography (New York). Fourteen books edited by her have been published by ICFAI University Press, Hyderabad. Her earlier publications include "Population Dynamics in India", "Industries and Environment: A Study of Impact Assessment", "Maharashtra" under States of Our Union Series by Publication Division, Government of India, and a text book on Geography for Maharashtra State Text Book Bureau, Pune.

Infrastructure Development and its Environmental Impact

Study of Konkan Railway

Prabha Shastri Ranade

CONCEPT PUBLISHING COMPANY, NEW DELHI-110059

Cataloging in Publication Data—DK
Courtesy: D.K. Agencies (P) Ltd. <docinfo@dkagencies.com>

Ranade, Prabha Shastri, 1950-
Infrastructure development and its environmental impact : study of Konkan Railway / Prabha Shastri Ranade.
p. cm.
Includes bibliographical references (p.).
Includes index.
ISBN 13: 9788180694509
ISBN 10: 818069450X

1. Konkan Railway. 2. Railroads—Environmental aspects—India—Konkan. 3. Railroads and state—India—Konkan. 4. Infrastructure (Economics)—India—Konkan. I. Title.

DDC 385.095 479 2 22

ISBN-13: 978-81-8069-450-9
ISBN-10: 81-8069-450-X

First Published 2009

Published and Printed by
Ashok Kumar Mittal
Concept Publishing Company
A/15-16, Commercial Block, Mohan Garden
New Delhi-110059 (INDIA)

Phones : 25351460, 25351794
Fax : 091-11-25357103
Email : publishing@conceptpub.com

Preface

Ever since the laying of the first ever railway line in the very bordering districts of the Konkan Region over 151 years ago and subsequent development of railway network in the Indian subcontinent during the pre-independence era, the Konkan Railway happens to be the first major railway project undertaken after Independence. For various reasons the implementation of this project was delayed. When the first phase of this project was cleared with green signal for commencing pre-engineering activities in the field, the local population had a sigh of relief and happiness with the hope of better future and the feeling of realization of their dream of having a railway line passing by their hamlets and through the countryside. A famous Marathi poet Vasant Bapat had expressed the very feelings of the commoners in a poem titled "Here comes the Konkan Railway", (*Aali Konkan Gadi*) composed some time during 1977-78. When such a long awaited project was successfully completed and was put into fully operational mode in 1998, it was considered to be a vital and most appropriate subject for undertaking the impact assessment study of the region, through which the Konkan railway is passing.

The results of the study were submitted in the form of a detailed report to University Grants Commission who had sponsored this research project work through Jawaharlal Nehru University, New Delhi Considering the importance and utility of the information contained in this report, which calls for wider circulation and easy access to the information. I ventured in publishing the same in a book form with due consent from the UGC. I am confident that this book will provide adequate and useful information to the academicians, students, researchers, economists, and planners and even to the social and political activists.

Well-planned and maintained infrastructure does have a positive impact on all-round development of the region. The findings of the study, elaborated in this book have shown a positive impact within a short span of its operations. It may take some time for visualization of major developmental and environmental impact generated, not only within the region but also in the neighbouring areas. The information contained in this book serves as database for conducting further assessment and studies.

Acknowledgements

The research project activities can proceed only when the preliminary information and relevant data base is made available from the various sources. My sincere thanks are due to Managing Director, PRO and other officials of Konkan Railway Corporation at Belapur, Ratnagiri and Madgaon, Central and State Pollution Control Boards, Maharashtra Remote Sensing Application Centre, Nagpur, Bureau of Economics and Statistics, Ministry of Environment and Forest, Maharashtra State Tourism Development Corporation, Department of Industries for providing valuable information. I express my deep sense of gratitude to Prof. Sudesh Nangia, Centre for the Study of Regional Development, Jawaharlal Nehru University, New Delhi not only for her guidance in my endeavour, in completing my project, but also for her help in getting this book published. I also thank M/s Concept Publishing Company for printing and publishing this book.

Contents

List of Figures

List of Tables

Introduction

Infrastructure and Development

The term Infrastructure relates to modern times and is associated with development and economic growth. Economic infrastructure is taken for granted in modern society. Infrastructure is the physical framework of facilities through which the services are provided for the social and industrial development. Infrastructure is an umbrella term for many activities referred to as "Social Overhead Capital". Infrastructure covers a complex of distinct sectors that represent a large share of an economy. Infrastructure, as defined by World Development Report 1994 on "Infrastructure for Development" includes the following services under the term:

- Power, telecommunications, water supply, sanitation and sewage, solid waste collection and disposal, piped gas, dam and canal works for irrigation and drainage, Transport sector roads, railways, ports, waterways and airports.

Infrastructure is divided into two components: Physical and Social. The sectors of transport, communication, irrigation, housing, water and power supply are grouped as "physical infrastructure." Social infrastructure consists of institutions and facilities "that would provide education, health and sanitation, banking and other forms of financial facilities and similar prerequisites to decent living. It is an integral part of the meeting of basic human needs. The ultimate goal of infrastructure policy is to provide the necessary services of high quality and efficiency at low price to the ultimate users in the country.

Infrastructure provides services that are vital to development. Improving economic infrastructure services enhances welfare and fosters economic growth. Provision and maintenance of adequate infrastructure facilities at reasonable cost is essential for achieving rapid and sustainable economic growth. Infrastructure being a *sinequanon* of economic development, its development is not a luxury but a necessity. Infrastructure represents the wheels of economic activity. In the absence of adequate infrastructure the benefits of economic reforms and socio-economic development cannot reach the masses. The adequacy of infrastructure helps to determine a country's success or failure in diversifying production, expanding trade, coping with population growth, reducing poverty or improving environmental conditions. The extent of penetration of a country's trade in the international market depends on the quality and efficiency of infrastructure. Infrastructure needs to match new demands as developing countries become more closely integrated into the global economy. Infrastructure can deliver major benefits in economic growth, poverty alleviation and environmental sustainability only when it provides services that respond to effective demand and does so efficiently. In order to achieve sustainable economic growth, developing countries across the world are focusing on providing basic infrastructure for their development. The availability of infrastructure has improved significantly in developing countries over recent years. India's economic reforms have necessitated the expansion of infrastructure.

Infrastructure and Environment

One of the major concerns in development projects is the environmental protection. Provision of infrastructure is directly related to environmental conditions. Agenda 21 of the primary policy document issued by countries at the 1992 Rio Conference on environment and development recognized the importance of infrastructure in maintaining higher environmental standards. Infrastructure has an impact on physical and socio-cultural aspects and has strong links with environmental sustainability. Environment friendly infrastructure services are essential for improving economic conditions and the living standards of the society and thereby of the country. Well-planned infrastructure helps to promote the environmental sustainability ultimately benefiting the human settlements. The relationship between each infrastructure sector and the environment is complex. Each sector—water, power, transport and irrigation raises issues concerning the interaction between man-made structures and the activities they generate and the natural environment. Railways and highways in general are no doubt an assault on the environment, but their development has become inevitable components of human civilization. Several types of negative impacts upon environment do happen during their construction, primarily due to negligent practices. The environment improves because of the existence of adequate transport infrastructure. Improvement in transport can contribute to growth which is environmentally sustainable. Better management of infrastructure provides environmental benefits. Thus infrastructure can be seen as a way of reducing adverse environmental impacts and a way of improving the health and standard of living of residents.

Public environmental awareness as such is a late twentieth century phenomenon. It has arisen as a consequence of increased level of pollution and the unsustainable use of resources in energy, transport and industrial processes. Awareness of the fragility of the earth's natural resources has led to mandatory requirements for conducting environmental impact studies prior to planning and implementation of all types of projects including those providing transport infrastructure. Environment of a nation is a natural legacy, which needs to be guarded and preserved. It is a subject specified in the Seventh Schedule of the Indian Constitution as the exclusive privilege of the Union Government. The constituent States cannot violate any law or regulation issued by the Union Government. It was after the Stockholm Conference in 1972 that India recognized the need for formulating more comprehensive laws for protection of environment. The concept of Environmental Impact Assessment for the projects was introduced in our country in 1989 when the Planning Commission issued instructions that environmental studies should be done for all projects taken up in the Eighth Five Year Plan.

Transportation: A Basic Component of Infrastructure

Transportation is a sector of major concern at the core of sustainable development. The present way of living is characterized by manifold and diverse transportation needs. Transport answers to people's desire to participate in different activities in different places. Availability of transport is considered as an index of economic, social and commercial progress. Roads, railways, ports and airports, the constituents of transportation, are vital links in the logistic chain that enable rapid movement of factors of production and distribution of goods and services. The transport industry has been profoundly affected by the developments in world economy and establishment of WTO. Competitive global trade requires reliable and sophisticated transport. Privatization has recently reformed the transport industry in our country. The role of transport in the national economy is gaining greater significance. The services associated with the use of infrastructure (measured in terms of value added) account for 7 to 11 per cent of GDP with transport being the largest sector. Transport GDP has grown 4.74 per cent per year during the last decade. Transport and communications constitute a major economic generator of activity in their own right. Therefore, the positive aspects of this (cultural exchange, financial returns) deserve to be strongly promoted.

Environmental Problems Associated with Transportation

Any highway or railway during construction disturbs the environment. However, the railway is recognized the world over as the most environmental friendly and far more energy efficient over other transport modes. The presence of a railway and its operation inevitably has an impact on the neighbouring areas. One is the nuisance created by the operation of railways and another is the visual impact resulting from the very presence of the railway. Caution has to be exercised in making an environmental issue of every activity intended to promote the welfare of common man. The development must proceed.

Use of land: Land and other resources are affected by transport infrastructure development. Transport uses land and is a major and a disruptive physical presence in towns and countryside. It is the lay of the land that gets disturbed when a new transport line is built. Transport route, therefore, seems to have a potentially large impact on land use. The expansion of rail and marine transport in the past and of road and air transport in recent years has consumed much of land both for new tracks and highways around ports and airports. In the cities of developed countries road surface alone normally cover 25 to 35 per cent of land. The main resource affected by transport systems is land. It has been calculated that surface transport systems (roads and railways) occupy 3.1 per cent of the total area and 4.6 per cent of usable land in our country. Roads occupy more land area than the railways. The road network occupies 18 times more area than the rail network. The extreme congestion on roads and severe problems with land acquisition for roads are unique to India, that gives rail mode, with its potential for rapid expansion of freight hauling capacity an added advantage. Railways are limited access divided expressways within partially fenced right of ways, they have the least problem with strip development and capacity expansion through land acquisition.

Ecological degradation: The land modes of transport require dedicated guide ways, the construction of which causes diversion of land use, which has environmental implications. Apart from this diversion construction of roadways and railways give rise to serious negative environmental externalities disturbing the layout of the land, drainage and watersheds by interfering with the free flow of water across the embankments. Embankments are raised but frequently the free flow of water across the embankment is obstructed. The construction sometimes cuts through natural forest or jungle. Trees are cut and vegetation destroyed when a highway or railway is constructed. This is against all canons of environmental protection. The railways in two ways cause deforestation: use of wooden sleepers and cutting of trees across the route. Wooden sleepers are totally avoided and concrete sleepers are used in the construction of Konkan railway.

Noise: Noise is one of the largest problems at present. It is one of the most widely recognized and resented environmental consequences of the increase in road and air transport. It interferes with work, prevents sleep, has psychological effects and can even damage hearing. The problem is especially serious in the town centers of developing countries where a variety of vehicles crowd the streets. Private cars generate sixteen times higher relative noise intensity than light rail transit (LRT) and buses four times more than LRT—for the same number of passengers transported. The special feature of trains is their intermittent passing compared with steady streams of road traffic. At night when freight trains operate on more crowded mixed traffic lines, the level and duration of their noise may be sufficient to interrupt sleep. Traffic management systems help to reduce noise and air pollution caused by unnecessary stopping and starting. In heavily populated areas, restricting train speed may be a viable means of noise reduction. Noise pollution created by railway system is not significant. Noise level measurements were well within limits during the construction stage of Konkan railway.

Energy consumption: The consideration of conservation of energy resources requires development of public mode of transport, specially the railways. Energy consumption is of utmost importance for the sustainable development of the country. Transport is one of the main energy consuming sectors in India. Almost half of the total petroleum products in the country are used in the form of high-speed diesel and gasoline in the transport sector. Several studies have shown that lesser traffic congestion, better roads and bridges infrastructure can save scarce petroleum products and bring about cleaner environment. In 1990 transport sector was the second largest consumer of commercial energy (22% of total) after industrial users (53%). Indian Railways accounted for only 4.3 per cent, while 18 per cent was used in road transport. Several studies show that railways are most energy efficient when performing services for which they use all resources efficiently. Railways and coastal shipping are substantially more energy efficient and environmental friendly compared to Highway transportation and private motorcars. For freight railways use energy most efficiently. Needs for minerals in railway construction are relatively modest. Less construction material is used in building railways because of their narrow width. Rolling stock of railways has relatively long life and can be intensively used. This is an advantage for railways. The low rolling friction of the rail and the higher power of the diesel engine allows rail roads to carry a large volume of traffic at a low loose mile cost.

Environmental pollution: There is scarcely any aspect of the environment that is not affected in someway by transport. Transport has a far wider impact on society than can be summarized under the usual list of pollutants. It has a major impact on health. Transport is a major contributor to environmental pollution as it requires the burning of fossil fuels to produce commercial energy. The transport sector contributes 30 per cent of all greenhouse gas emissions. The emission of gases from vehicles contribute to global warming and are a major influence on local air quality. The main constituents are carbon monoxide, lead, nitrogen oxide and hydro carbon. The rate of pollutant emission per vehicle is much higher in India than in other countries. Aircraft and railway locomotive together emit a far smaller volume of air pollutants than road vehicles do. The highways with high capacity trucks cause the maximum land and environmental abuses. Large cities in the region face the problem of air pollution due to increase in number of vehicles. Road transport causes more pollution through the exhausts of millions of vehicles which are less efficient than trains. Not only the traffic police but the pedestrians, daily commuters, shop keepers and common men are affected by this high level of pollution. Urban air pollution studies in India have revealed that vehicular pollution is the predominant contributor to air pollution in mega cities. 63 per cent of air pollution in Mumbai is attributed to vehicular pollution. Running trains indirectly contribute to pollution: (1) The toilet discharges by passengers using train; (2) Accidental spills from the goods trains; (3) Sewage from railway stations, public toilets, railway staff colonies. There are chances of contamination of water if these discharges find way to surface water sources, and this may affect public health. Such discharge points are constantly monitored for avoiding any kind of contamination.

Public safety: Roads in India are dangerous by developed country standards with an annual fatal accident rate of about 2.65 deaths per 1000 registered vehicles. WHO estimates that every year more than a quarter million people die in road accidents world-over, about 60,000 among them in India alone. In this respect India is at least six times worse than the worst of the European countries. The roads are in lead again, since private cars are involved in traffic accidents twenty times more frequently than LRT and ten times more frequently than city buses. Unauthorised occupations pose a major traffic hazard to the road users as well as to the roadside community. There is an increase in number of vehicles involved in road accidents, number of persons killed and injured. Comparing road and rail on the parameter of death, rail is a distinctly safer mode than road, though there are several level crossing accidents and deaths due to illegal padestrian crossing rail tracks. Safety is a matter of concern for Indian Railways. In India railway as means of public transport offers relatively higher security. The Railways have passenger fatalities due to train accidents of only 100 per year despite the fact that they account for approximately 16 per cent of passenger km nationwide. Thus in addition to being more benign environmentally, rail transport is clearly the safest mode of passenger transport in the country.

Railways: An Environmental Friendly Mode of Transport

Railway is recognized the world over as the most environmental friendly, energy efficient and one of the least polluting among the modes of transport. Railway is the most efficient in the land based transport sector whether the criterion is land use or energy consumption or pollution by emission and noise. It is the railways that inflict the least adverse influence upon the environment and may thus be viewed as eco-friendly. Rail industry contributes meaningfully to achieving desirable environmental, economic and social goals (cutting pollution and congestion). The advantage of railway system is that it can transport large number of passengers and huge quantities of goods in fastest possible time leading to cost advantage. Electric traction which does not cause pollution has already been introduced to cover high density routes. In the modern world passenger train has been reinvented and reinvigorated as a meaningful contributor to the mobility of the countries they serve. High speed trains are now attracting air travellers because of low travelling costs and comforts. In China, India, Russia and Southern Africa new lines are still being constructed and planned; elsewhere existing lines are being upgraded for modern rolling stock. Terrain, long haul distances and the distribution of population and economic activity in India all support a much larger potential role for railways than in most countries. With the growing awareness of protecting the environment, concern for quality of human life and for conserving the limited energy resources, India has no option but to turn increasingly to using the railway intensively or else have the coming generation pay the penalty for our lack of far-sightedness. In the transport sector railways are the driving force for India's economic progress. Indian railways have completed over 150 years of existence and operation in the country. Indian railways epitomize the nation's economic development. There is still scope for intensifying or accelerating train services. Commissioning of Konkan Railway (KR) in January 1998 was a major step towards infrastructure development in India in transportation sector.

The Indian Transport Scenario: Emerging Trends

Though the domestic mechanized transport system in India consists of railways, road services, coastal shipping, inland waterways and airways, railways and roadways are the two most important inland transport services in the country. Terrain, long haul distances, the distribution of population and economic activity in India all support a much larger potential role for railways than in most countries. India's transport infrastructure does have an inherent strength due to its extensive reach and variety. Its infrastructure base consists of the third largest road network in the world, 2.1 million. km., second largest railway network of over 63,000 km route length and 12 major and 139 other ports. Historically the responsibility of providing transport infrastructure has been vested with the government. Road transport in India is a dominant mode of transport both for the movement of passengers and goods. In keeping with the economic development in the country, road transport has experienced substantial growth over the years. The traffic volume moved by road has grown rapidly compared to those moved by rail. Roads have emerged as the preferred mode of transport in comparison to rail because of certain inherent advantages. This has been changing the rail road mix in transport sector continuously in favour of road. In the Indian transport sector there are three important historical developments to bear in mind:

(1) India, a rail dominated economy in the 1950s has become a decidedly road dominant economy in the 1990s. The share of road for freight traffic has increased from 11 per cent in 1950-51 to 60 per cent in 1994-95. Similarly, the share of passenger traffic has increased from 28 per cent to 80 per cent.
(2) Indian railway shifted from being a freight dominant operation to a passenger dominant operation.
(3) The main links of the parallel and competing road and rail network have become saturated under the current technological and operational regimes.

The traffic by road is likely to increase substantially in the next decade due to countrywide development, second generation economic reforms and the completion of on going Golden Quadrilateral Four-lane Highway project. The economic reforms initiative in India is expected to raise the annual growth rate of the Indian economy. This implies a doubling of freight transport output every 10 to 13 years and a doubling of passenger transport output every 7 to 10 years, as the demand for transport is closely linked to measures of economic growth. This situation calls for expansion of transport infrastructure. National Transport Policy Committee and Railway Reforms Committee have recommended a clear cut policy for the construction of new railway lines in the country for balanced socio-economic development of the country. The order of priorities for this agreed by the Ministry of Railways is as follows—(i) The project oriented lines, (ii) Lines to serve as missing links, (iii) Strategic lines, and (iv) Developmental lines. Konkan railway is a broad gauge single line from Roha in Raigad district of Maharashtra, (145 km from Mumbai) to Thokur near Mangalore in Karnataka along the western coast. Konkan Railway serves as a missing link between Mumbai and Mangalore. An observation of the transport map shows that completion of this missing rail link provides the much needed longitudinal artery through the west coast. It can also be called as strategic line and developmental line.

Objectives of Study

The environmental consequences of a project are commonly referred to as "impacts". Any development project is bound to have some impact on the environment. Transport projects have direct physical impacts on the environment. The indirect impacts are normally socio-economic and cultural in nature. With the initiation of construction activities of a project, a chain of events follows and a series of impacts are triggered by various activities during construction, upon completion and during its operation.

This project aims to analyze the impact of major infrastructure project of Konkan Railway on the environment and socio-economic profile of the western coastal region of India. It intends to study the spatio-temporal pattern of change by incorporating the study of resource base, population growth, urbanization and industrialization, passengers and goods transportation, environmental conservation and operational safety measures and rehabilitation measures taken by Konkan Railway and their impact on adjoining region. To be precise, this is a study of the impact of Konkan Railway on:

(1) The environment and land use,
(2) Population growth—rural and urban,

(3) Tourism development,
(4) Investment in infrastructure and industry, and
(5) Major ports on western coast connected by Konkan railway.

The impact of infrastructure on regional economic development has been found to be highly significant and positive. Infrastructure development provides an impetus for population growth and urbanization of the region. Transport corridors attract industrial development and influence the land use. Rapid industrialization, and agricultural development together increase freight and passenger transport and higher real incomes stimulate leisure related travel. The major impact of modern transport on the environment is visible in the growth of tourism. The positive contribution of infrastructure to economic growth and development comes through increases in investment, employment, output and income in a chain of "cumulative causation." Thus "economies of agglomeration" develop over time leading to further concentration of economic activities in a particular region. The impact of Konkan Railway is assessed on the background of above observations. World Development Report on infrastructure observes that "Infrastructure development leads to environmental improvement." Infrastructure development provides an impetus for further socio-economic development of the region. How far the Konkan railway meets the challenge of achieving sustainable development while protecting and improving the environment is examined.

Organization of Study

The present study is divided into nine chapters. The first chapter deals with the spatial structure of the Konkan region. It describes the geographical setting and the environment around Konkan Railway, distribution of forests and transport infrastructure of the region. Chapter 2 presents the profile of the Konkan railway project emphasizing its technological innovations and new work culture and gives an account of passengers and goods transportation. Chapter 3 on land use changes gives an account of land acquisition for KR, land use along the KR alignment and within the 20 km corridor (10 km on both sides) at the time of acquisition and within districts through which KR is constructed, loss of agricultural land and the impact of KR construction on various categories of land use including forest. Chapter 4 includes an account of population displacement due to land acquisition for KR, rural population growth, emergence of new towns and growth of towns and stations located along KR. Chapter 5 considers the role of KR in tourism development. Impact of KR on attracting both domestic and foreign direct investment for infrastructure and industry and subsequent industrial development is studied in chapter 6. Chapter 7 discusses the impact of infrastructure development and improved connectivity on major ports of western coast connected by KR. Chapter 8 gives the environmental conservation measures including the afforestation activities, operational safety measures on KR, measures taken for resettlement and rehabilitation, employment to land losers and employment generation by KR. The last chapter summarizes the impacts and benefits of Konkan railway line and evaluates the role of KR as a sustainable transport infrastructure and an asset for regional development.

Data Base

This study is based mainly on secondary data compiled from the latest and past publications of the Census of India, Directorate of Statistics of respective State Governments, National Remote Sensing Agency, Forest Survey of India, Remote Sensing Application Centre, Central and State Pollution Control Boards, Konkan Railway Corporation and other agencies. Relevant information, maps and unpublished statistics were collected by personally visiting various data generating agencies, including the Corporate and zonal offices of Konkan Railway Corporation at Belapur, Ratnagiri and Madgaon, discussing with the concerned officials and were updated from various websites. District and tahsils are the units of analysis of secondary data. The situation prior to construction of Konkan railway as described in past records is compared with the present situation based on the latest available statistics in order to assess the impact of each variable.

Relevance of the Study

In order to accelerate the pace of growth and to sustain high levels of productivity in all sectors of economy, infrastructure development is accorded high priority in India's development strategy. In our country the demand for infrastructure is ever increasing. Higher economic growth and urbanization are bound to generate accelerated demand for infrastructure facilities. The CII's National Conference in 1997 pointed out that India's transport infrastructure has desperately failed to keep pace with the demand of the growing economy. In recent years as a result of higher economic growth, the development of transport infrastructure in particular has acquired a sense of urgency. The key issue is that the country cannot achieve 7 to 8 per cent growth rate unless there is a substantial expansion and enhanced efficiency in all areas of infrastructure.

New railway line construction is a challenging task as it is followed by enormous social and economic benefits. On the background of increase in environmental pollution generated in the transport sector, any step towards environmental improvement is highly desirable and welcome. And when this improvement is associated with a new railway line construction which had a high demand, it assumes added significance. KR is the biggest project of a new railway line construction undertaken on the Indian subcontinent in the 20th century. Konkan railway is a unique railway system in the world because of its length of over 760 km running more or less parallel to the coast through an inhospitable terrain. Similar situation does not prevail in other parts of the world. Notable cases of coastline railways worldwide are on the Spanish, French and Italian rivieras and the coast of Britain. In Sri Lanka a railway runs on or near the shoreline for 170 km. On this background, the length of Konkan railway is appreciable.

Several development projects were undertaken in our country after Independence under the Five Year Plans. Large number of studies are available on urban transport/infrastructure problems. Impact of public sector enterprises has also been the focus of some studies. Many researchers have examined the impacts of new public transport systems around the world, and their benefits. Though studies pertaining to impact of a variety of infrastructure projects including highways, irrigation, electrification, ports are available, they mainly emphasize on the socio-economic impact. A study of environmental impact of railways published in Britain informs railway planners about the environmental impacts which they may cause and suggests how they can design optimal solutions. Two publications brought out by the Konkan Railway Corporation give the background and technical

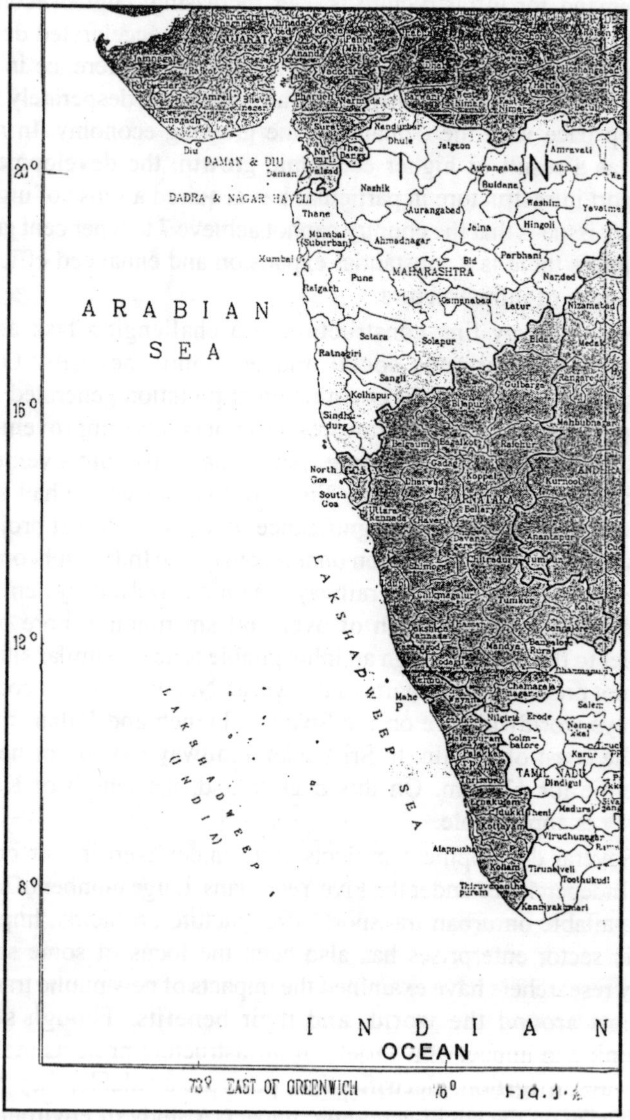

Based upon Survey of India map with the permission of the Surveyor General of India. The territorial waters of India extend into the sea to a distance of twelve nautical miles measured from the appropriate base line.

Fig. 1.1 Western Coastal Region

details of the railway line and serve as a guide for future construction engineers. The aspect of environmental impact of an infrastructure project after its completion has received little attention in the academic literature. The impact of HBJ gas pipeline, an infrastructure project, on the environment of the adjoining state was examined by this researcher. However, such a study of environmental impact of a newly constructed transport infrastructure project of the magnitude of Konkan railway for its entire stretch has not been undertaken by anyone so far. The impact of Konkan railway on various aspects of regional development and the spatio temporal pattern of change has also not been assessed so far. The present study is thus pioneering in this area, relevant and timely. It attempts to make an original contribution in the field of Transport Geography in a region, which remained underdeveloped due to lack of infrastructure.

Areal and Temporal Framework

The Western Coast of India comprises the coastal lowlands of Gujarat, Maharashtra, Goa, Karnataka and Kerala. The Konkan is a narrow coastal strip of land bounded by the Arabian Sea in the west and the Sahyadri mountain range (Western Ghats) in the east, which give it a distinct regional identity. The present study is confined to the coastal districts through which the Konkan railway is passing, namely, Raigad, Ratnagiri and Sindhudurg districts of Maharashtra, North Goa and South Goa districts of Goa state and Uttar Kannada and Dakshina Kannada (including Udupi) districts of Karnataka (Fig. 1.1). The region spreads over an area of 42,636 km^2 and has population of over one crore (10,467,000) in 2001. The region has a long indented coastline of about 850 km (Maharashtra 480 km, Goa 130 km and Karnataka 240 km). Despite the western coastal region's rich resource base, poor infrastructure conditions was a major constraint to its economic growth. There are intra-regional variations among the northern (Maharashtra), Central (Goa) and southern part (Karnataka) of this region in respect of the natural landscape, resource endowment and levels of development. Historical and political factors too have played their part in promoting or retarding the development of this region, e.g. Goa as a Portuguese territory until 1960, Dakshina Kannada district under Madras Province and Uttar Kannada district under Bombay Province prior to reorganization of states. Study of the impact of Konkan railway on environment of this region cannot be treated in isolation, from these events.

The hypothesis of expansion is that a strong demand exists in the new area chosen for laying the railway track. Konkan railway meets the long standing demand of infrastructure for India's western coastal region. The first survey on the Konkan railway route was conducted from Kalyan to Mahad in 1866 and several surveys were conducted subsequently. Konkan railway was a long cherished dream of the people of Konkan region and governments of coastal states for over a hundred years ever since Bombay (Mumbai) was being developed as a major industrial centre and a commercial capital of the country. Bombay offered employment opportunities for the people of Konkan region and in the absence of dependable mode of transport (railway) people of the west coast had to travel either by steamer service or by buses. Demand of this railway line construction became more persistent as the years progressed after independence.

In order to execute this project, the Konkan Railway Corporation Limited in (KRCL), a separate statutory body was set up on January 1990. The construction of the project commenced in October 1990. The project was opened up in phases from either ends progressively. From the southern end Mangalore-Udupi section of 67 km was opened on 20 March, 1993 and from the northern end Roha-Veer section of 48 km was opened on 27 September, 1993. The entire stretch became operational since 26 January, 1998. This study covers time span of over a decade starting from pre-construction period in 1989-90 till its completion and full length operation up to the year 2001-2002.

1

The Konkan Region: Spatial Structure

Geographical Setting and the Environment Around Konkan Railway

Konkan is a narrow coastal strip of land bounded by the Arabian Sea in the west and the Sahyadri mountain range (Western Ghats) in the east. The most important topographic feature of peninsular India is the 1400 km long Western Ghats along its western margin spread over in five southern states almost parallel to the west coast and often hardly 40 km from it. The Konkan railway alignment literally carved out of the Western Ghats passes through one of the most difficult and formidable terrain in India both physically and geologically with nearly 1500 streams to be crossed. The Western Ghats are the main water shed in peninsular India from which all principal rivers originate and flow towards east and west. The geography of Konkan region is typical for its irregular landscape and rugged and hilly nature of terrain. In Konkan streams originating on the Sahyadrian scarp meander their way to the Arabian Sea traversing the coastal plain flow from east to west. Deep valleys of east-west flowing rivers which have deep gorges in the hills and wide creeks as they approach the sea, carve the intermediate landmass. In-between the rivers, are high rising ridges or plateaus with elevations rising up to 200 m above mean sea level. In Goa as well the situation is more or less similar, although there are some portions where there is plain land between east-west rivers. In coastal Karnataka there are plain lands near the sea coast. Thereafter one encounters the foothills of the Sahyadris towards the east. In the plain land there are tidal rivers while in the foothills of Sahyadris the rivers flow through gorges or deep valleys. The bed level of the rivers is very close to sea level and the tidal range reach up to about 40 to 50 km from the sea. The construction of a railway through such a region posed formidable problems.

The Konkan Region

Maharashtra Coast

In Maharashtra the terrain consists mostly of plateau lands or rocky hills, with innumerable and irregular spurs and offshoots intersected by east west rivers.

Raigad District

Commencing from north to south Raigad district forms an important part of the Konkan plain. It begins from the boundary of New Bombay (Navi Mumbai) and the Kashedi mountain range forms its southern boundary. Raigad district is slightly elongated in the north-south direction with an area of 7148 sq. km. Raigad has a long indented coastline with a number of creeks and inlets, suggesting submergence. Mangroves, marshes and tidal flats are common specially in the north. Ruggedness and uneven topography form the governing theme in its physical features. About 100 km south of Bombay, the Panvel-Pen-Alibag coastal plain gives place to a more rugged topography and a general rise in the height of Konkan. The Sahyadris in the east form a long unbroken range within the district. It is well forested and rises in places to more than 500 metres. The Sahyadris send several transverse members of subsidiary hills westwards denying the plains of a uniform level and continuous character. The hills are well forested, being mainly reserved forests. The region is characterized by steep scarp slopes. The rivers appear entrenched and a height of 200 m of the flat interfluves gives the aspect of a coastal plateau. Important rivers in this district are Kal, Patalganga, Amba, Kundalika. Savitri river forms a boundary between Raigad and Ratnagiri districts to some extent. The Kal river is tidal for a considerable distance. The plateau in between the Sahyadris and the coastal plain is of about 100 metres though there are spot heights which rise to more than 100 metres. Forest cover is negligible over the plateau and limited in Kal-Savitri Valley. Small pockets of reserved forests occur in Murud and Shrivardhan tahsils of Raigad district. A five hectare isolated patch

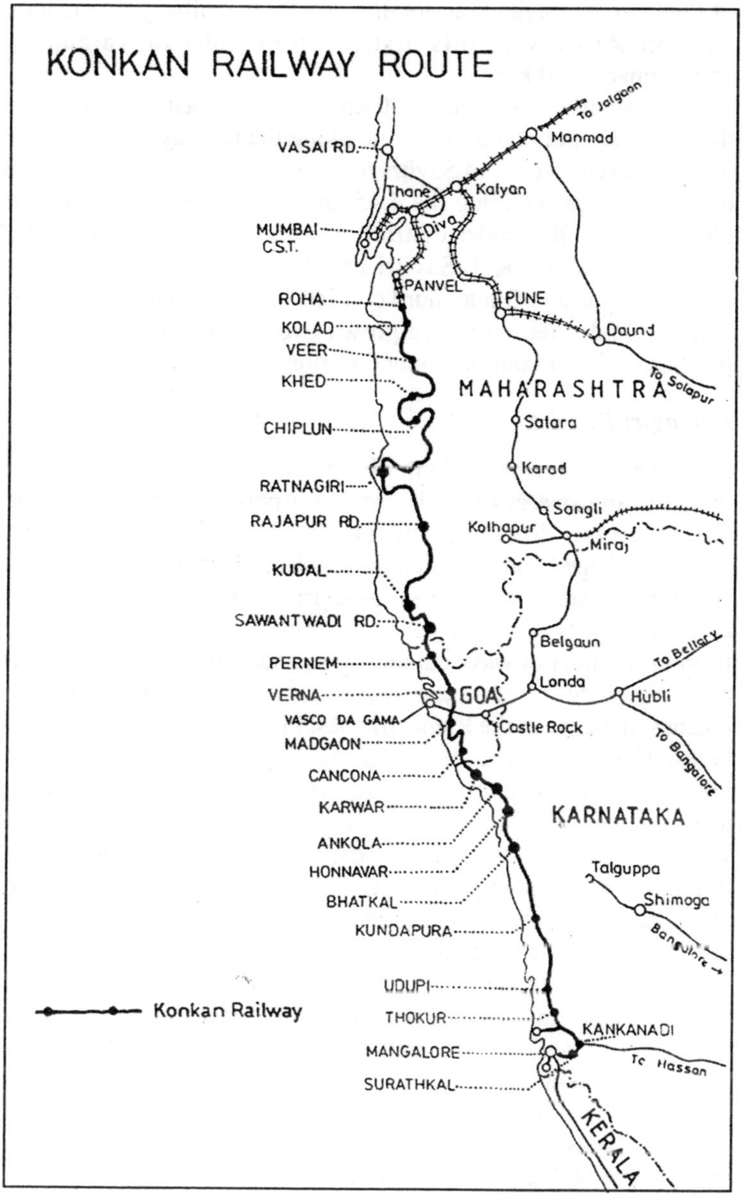

Fig. 1.2 Konkan Railway Route

of forest in mountains was got diverted for the railway. The length of Konkan Railway in this district from Roha to Karanjadi is approximately 70 km.

The alignment lies about 30 km from the coast and 20 km from the Sahyadri mountains. The land along the railway consists mainly of plains except south of Savitri river, where it has to cross a number of ridges before entering the 4.45 km long tunnel through Kashedi ghat to enter Ratnagiri district. In this stretch it crosses Patli, Kundalika, Ghod, Kal, Savitri and Nageshwari rivers. The area between Roha and Ghotekhurd consists of fertile plains. Most of it is irrigated and three crops are grown every year. There are isolated patches of forest land in mountains along the railway.

Ratnagiri District

Ratnagiri district has a long elongated shape in the north south direction and an area of 13,054 sq. km. Approximately 85 per cent of the land in this district is hilly. For 16 to 24 km from the coast the terrain is rugged and mountainous. Unlike the coast of Maharashtra which is generally low lying, Ratnagiri coast of a long narrow strip of land is cliffy. Erosion seems to be predominant as can be seen from the number of wave cut platforms and cliffs. Beaches in general are not well developed and are very narrow and occur only in small patches. In between the Ratnagiri coast in the west (which is generally below 100 metres) and the Sahyadri hills in the east (which is above 200 metres) lies an elongated plateau region. Within the region there are a number of spot heights that rise to more than 200 metres. The Sahyadri hills in this district are characterized by very steep slopes. A number of isolated peaks, located in the region have a height of more than 200 metres. The highest spot height is 1239 metres in Khed tahsil and its surrounding region is covered with dense forest. The rivers in Ratnagiri district are short and swift and though Ratnagiri receives heavy rainfall, the rivers dry up to mere trickles soon after the monsoon season. The important rivers are Jagbudi, Vashishti Shastri, Muchkundi, Kodavali. These rivers dissect the plateau which, in southern part of Konkan is covered with laterite. All the rivers are tidal for a considerable extent. The creeks of these rivers are of little value from navigation and fisheries point of view. The westward flowing parallel streams bring vast quantities of eroded material and deposit it in the lowest zone that abuts the shoreline. Silting of rivers and formation of sandbars near the confluence of these rivers with the sea are a continuous and normal occurrence. Forest cover is limited in Ratnagiri district and is mainly in the southern part of the region.

The length of Konkan railway in Ratnagiri district is 210 km. Here the line passes through one of the most rugged and difficult terrains along the route and it is this section that has the maximum number of bridges, with high viaducts and long tunnels. The railway alignment follows the valleys of rivers flowing north-south and crosses the plateau with tunnels. The rivers flow almost at sea level and it is difficult to gain height to cross the plateaus within the limitations imposed by the ruling gradient of the railway line. The alignment crosses Jagbudi river and follows its valley upto Chiplun. Here it crosses Vashisti river. After passing through a series of tunnels, it crosses Shastri river. Thereafter the alignment rises to reach Ratnagiri plateau through a number of tunnels, the longest being the 6.5 km long Karbude tunnel. After Ratnagiri the line has to descend to cross the Kajli river. This is achieved by an easterly swing for about 20 km. Thereafter the rivers Muchkundi, Rajapur and Vagothan are crossed to reach the southern border of the district. Most of the land is covered with degraded forest and mango orchards. The famous alphonso mangoes are grown in this area.

Sindhudurg District

Sindhudurg district has an area of 5207 sq. km and a coastline of 121 km. The outstanding feature in the relief of Sindhudurg district is its highly uneven nature and the very narrow riverine plains that fringe the coastline. The area to the north is hilly and rugged but the district has some flat surfaces too. South of Kankavli the land is generally less hilly, interspersed with river valleys. The important rivers in the district are Devgad, Piyali, Janoli, Kasal, Karli Achara, Vaghothan river forms its northern boundary with Ratnagiri district and Terekhol river forms its southern boundary with Goa. The creeks of these rivers are of little value from navigation and fisheries point of view. Extensive spreads of laterite, the predominant soils in the district are noticed throughout. Due to the inundation of the sea, a part of the coastal soil has become salty. They are locally known as Khar or Khazan. In Devgad, Malwan and Vengurla tahsils the entire western strips are salty. The coastal strips have deep sandy loams. The estuarine portions have good soils and they form agriculturally the best lands of the district. The district is famous for its mango crop, the world renowned Alphonso, coconut and arecanut gardens, fruit trees and spices. Area under forest is limited and constitutes 7.5 per cent of its geographical area. It is concentrated in the northern part.

The length of Konkan railway in this district is 102 km. After crossing the Vagothan river the railway alignment climbs the Vaibhavwadi plateau and cuts across two tunnels Vaibhavwadi and Kharepatan, then follows the National Highway and enters Goa after crossing the Terekhol river. A number of major and minor bridges are constructed along the alignment.

Goa

The state of Goa is located along the mid-west coast of India to the south of Maharashtra and to the north of Karnataka. Terekhol river divides Sindhudurg district and Goa. On the west is the Arabian sea and the coastline stretches to a length of about 105 km. The state covers an area of 3700 sq. km with the north-south length of around 96 km and width varying from 25 to 55 km. It has many physical features that are common to the neighbouring regions of Maharashtra and Karnataka. Goa is divided into four physiographic divisions. The western ghat hill ranges occupy about one-third of the total territory of Goa. The slopes of the Western Ghats are covered with forests. Most of these hill ranges are being mined. The central part of Goa or the Midland region is occupied by undulating uplands having gentle to moderate slopes. These are mainly under paddy cultivation. The western and south-western parts of Goa constitute the coastal plains. The coast line of Goa is thickly palm fringed. The flood plains of the two major rivers, Mandovi and Zuari divide the coastal plains and rolling upland. The Mandovi has the largest drainage basin in Goa. Both these rivers have extensive estuarine land below the high tide level and these lands are locally called as *khazan* or contour lands. Other rivers flowing westwards into the Arabian sea are Tirocol and Chapora. South of the Zuari river there are plains bounded on the east by mountains. The Sal river flows north to south till Bali and then enters the plains of Canacona.

The length of Konkan railway in Goa is 105 km. The alignment runs through generally hilly land between the rivers Terekhol and Chapora. Then it runs through the central portion of Mandovi and Zuari rivers and runs for about 30 km through the *Khazan* lands from Assnora-Mayem to Cortalim. Some *Khazan* lands are also found along the track near Madgaon and south of Chaudi (Canacona). South of the Zuari river there are plains bounded on the east by mountains. After crossing the Sal river the alignment reaches Bali where a chain of high hills is intercepted. To the south of these hills the alignment enters the plains of Canacona and here after crossing the two rivers—Talpona and Galgibag the line crosses a high plateau to Uttar Kannada district. Konkan railway passes through dense forests in Quepem and Canacona talukas through a length of about 7.5 km. Along the hill slopes there are a number of forest plantations like Eucalyptus and Teak.

Karnataka Coast

Immediately to the south of Goa lies Uttar Kannada. This district has a distinct coastal plain of 8 to 16 km width. The Karwar-Honavar coast is characterized by coastal landforms. It is indented by several creeks and estuaries formed by Kalinadi, Gangavali, Aghanashini, Tadri and Sharavati rivers. This estuarine plain is connected by narrow coastal strips, interspersed by hill ranges extending from the Western Ghats to the Arabian sea. The land is generally fertile and thickly populated. The Western Ghats extend over the central parts of Uttar Kannada district in the form of north-south running elongated zone. This entire hilly zone is made up of a series of parallel ranges intersected by several rivers and their tributaries flowing towards west. Rock outcrops are also extensively observed. Soils of the region are shallow, black, brown and alluvial. Large stretches of undulating land comprising half of its total area is under forests. To the east of the coastal plain lies the hilly region of varying heights of 250 to 500 m. It is an undulating terrain broken up by chains of rocky hills and scourged by deep ravines. The Sahyadri mountains, with an average height of about 885 m lie to the east of this region. Rounded hills and broader valley features mark the transition between the main Western Ghats landscape and the drier central Karnataka plateau. This region almost fully covered with thick evergreen reserved forests is also known as Malnad. Yellapur-Dandeli forest is in general, an extension of the Western Ghats region. Sirsi-Siddapur dissected tract extending over eastern parts of Uttar Kannada district represents quite a dissected and complex landscape. The entire tract has been badly dissected by rivers and their numerous tributaries. Tadri and Gangavali rivers originate from this tract. The soils are shallow black, brown and alluvial. It is largely covered with forests and plantations.

The length of Konkan railway in this district is about 133 km. The railway alignment passes through the coastal plains which are the most populated and where the bulk of commercial activity takes place. It runs almost parallel to National Highway 17.

Dakshina Kannada district, located to the south of Uttar Kannada district and north of Kerala forms a long and narrow strip along the coast. The coastline of Dakshina Kannada, approximately 135 km in length, is narrow and unprotected and comprise mainly sandy beaches. The Coondapur-Mangalore coast represents a relatively narrow belt of coastal lowlands of very recent deposits varying in width from 10 to 20 km. The coast line is indented by numerous bays and creeks which have been formed by the river estuaries. In elevation it is generally low with an average of 60 metres. It is also interspersed with sand bars and broken rugged rock outcrops at some places. The region has high base status soils of humid regions and recently formed alluvial soils. The Western Ghats in the northern part of the district assume the form of steep scarps, but in the south it has a character of parallel ridges intersected by deep valleys. The elevation reaches up to 1500 metres and relative relief is quite high. Most of the westward flowing rivers originate here. There are several peaks rising from 900 m to 1800 m. The most important of these are the Kudremukh group of peaks. There are several passes from the plains to the hinterland beyond the Sahyadri. The main passes are those connecting Byndoor to Someshwar, Udupi and Mangalore to Bettanwadi via Buntwal. Due to heavy rainfall there are several stretches of thick forests. They are dense and evergreen and provide the timber of very good quality. Beltangadi dissected zone in the southern part of the district represents a very complex and dissected landscape full of hills, valleys, ravines, and forests.

The Bainduru Bantval plain extends over central part of the district in the form of an elongated tract running north to south. It is a somewhat broken plain between coastal region and the Western Ghats. In general the elevation is about 100 metres. Its slope is towards west. It is intersected by numerous rivers, viz. Kollur, Chakra, Madisal, Swarna, Mulki, Gurpur, Netravati.

In this district the alignment of Konkan railway lies in the coastal strip which is long and narrow and consists of fertile plain land. It also runs almost parallel to NH 17. This region is thickly populated as it is fertile and has trading facilities and commercial and industrial activities. The railway alignment having a length of 140 km connects the main towns of the district situated in the coastal plain.

Geology

Geologically the Konkan lowlands are a platform of marine denudation raised to form a narrow plain. The steep scarps facing the coast and somewhat gentle slopes towards the east is believed to be due to extensive faulting of the ghat region. Geologically the entire west coast strip from Roha to Mangalore falls in two distinct zones. The first zone extends from Roha to Kankavali and the other zone extends from Kankavali to Mangalore. In the first zone geological formation consists of dark coloured volcanic lava flows and laterites. These are dominantly basaltic in nature. The geology of Raigad district consists of dark coloured volcanic lava flows and laterites which date back to the Mesozoic era. These are spread out in the form of horizontal sheets and beds and have innumerable spurs, hills, ridges peaks and plateaux. In the plains and valleys the lava flows are covered by a thin blanket of soil which varies in thickness from place to place. The geology of most of the Ratnagiri district is Deccan trap with intertrappean beds. These are dark coloured volcanic lava flows like Raigad district. In the south of the district the geology varies being laterites, Penganga beds, Pakhal and Kaladgi series, granites, granite Gneisses and Dharwar system Sausar and Sakoli series. The laterites dominate the landscape of south Konkan. Laterites are seen extensively on plateaus and slopes of every hill or spur in the area. The occurrence of laterites in south Konkan has resulted from heavy rainfall (2500 mm), leaching of the rocks and their hardening. The laterites are derived from basalt and are rich in iron and aluminium. The depth of laterite is variable, with a maximum thickness of about 10 m near Ratnagiri and about 20 m near Rajapur. They vary in colour from bright red to brownish red due to the preponderance of hydrated iron oxides. They are fairly well supplied with nitrogen and organic matter and their texture is loamy. They are porous and not retentive of moisture. The coastal plains of Goa are formed by the constructional landscapes of the Arabian sea. The formation of the two major river systems of Goa—Mandovi and Zuari reveals the cleavages that have formed from the faults during the process of upliftment of the landmasses. The main geological formations in coastal Karnataka are Dharwad, Archaean, gneisses, granites and recent alluvium, schists, charnockites and basic intrusives. The soils are recently formed alluvial, red loamy, red sandy, shallow black and brown. They are fertile and suitable for cultivation. The entire south Kannada district is represented by Dharwar system of rocks and some non-dated granites and Charnokites. The coast is of recent geological origin and is still in an unstable stage. Coastal erosion is a major problem due to unprotected coastline and significant wave force impacts, specially during the monsoon.

Minerals

The distribution of minerals is closely related to the underlying geological formations. The Deccan trap generally lacks minerals of economic importance, but being hard, dense and durable are most suitable as building material. Konkan claims some of the useful deposits of minerals in Maharashtra. Ratnagiri and Sindhudurg districts seem to be favourably placed in regard to the availability of some of the important minerals. Bauxite and ilmenite are found at a few places along the coast near Harnai, Vijaydurg, Devgad and at Amboli in Sawantwadi tahsil. The deposits near the coast are important from the exploitation point of view. Nodular kankar, a concretionary lime carbonate is found sporadically on soils and alluvium covering the Deccan trap in Raigad district. In addition to kankar, the sea shells found along the coast at places are used for manufacture of lime. Salt is produced by direct solar evaporation of sea water near Uran, Sheva, Pen and Panvel. Reserves of iron ore are found around Redi, Banda, Tirwade and Oras in Sindhudurg district. The proportion of ore carrying between 58 to 60 per cent of iron is found particularly in Redi. Ratnagiri and Sindhudurg districts also have the deposits of silica sand, chromite and china clay. The reserves of chromite in Ratnagiri district are found in the beach sands stretching about 40 km from Purnagad to Malgund near Ratnagiri. Reserves of chromite and silica are also found in Kankavli tahsil. Manganese is found in Sindhudurg district at Phondey, Sasoli, Netarde and Dingne. Other minor minerals found in Ratnagiri and Sindhudurg districts are feldspar, quartz, mica, clay, asbestos, building stones, glass sand, limestone, etc. Deposits of clay are found in Rajapur, Dapoli and Chiplun tahsils. It is used for making tiles, bricks and pottery. Uranium oxide is also found in some parts of the district.

The coastal districts of Karnataka are outside the great flow of traps. In Uttar Kannada district there are deposits of Manganese ore, iron ore, limestone, silica sand and clay. The iron ore occurs in the western half of the district in the tahsils of Ankola, Honavar and Yellapur. There are extensive reserves in Apsarkond area of Honavar tahsil, Bisgod, Kalche and Kadlagadde areas of Yellapur tahsil and Kuntagani area of Ankola tahsil. Iron content in these deposits ranges from 58 to 65 per cent Manganese ore of this district is characterized by a high manganese content coupled with low phosphorous and low electromagnetic elements. Manganese bearing areas are Supa,

Yellapur, Sirsi, Kumta, Ankola. Limestone occurs in large quantities in the tahsils of Supa, Yellapur and Kumta. Most important clay deposits in the district are at Castle rock and Kuntagani area of Supa tahsil, Achave in Ankola and Heravatta area of Kumta tahsil. Bauxite deposits are at Mungolli and Talgod of Bhatkal tahsil. There are copper deposits near Kaiga, glass sand near Bengere area and ilmenite sand at numerous places all along the sea shore. Granite and laterite are very useful as building material. The South Kannada district is not rich in mineral wealth but contains some useful minerals such as bauxite, banded granite, soapstone, corundum, garnet, iron ore, silica sand, lime shell and clay suitable for the manufacture of tiles in considerable quantities.

The state of Goa is rich in minerals such as iron ore, manganese ore, bauxite and silica sand. Mining is an important activity for the economy of Goa and a significant foreign exchange earner for the country. Goa has been endowed with certain natural factors like a good natural harbour and navigable perennial rivers that facilitate the economic exploitation of these mineral deposits. The mining belt of Goa extends over 700 sq. km. in Bicholim, Satari, Sanguem and Quepem. Mining activity has led to major environmental problems like siltation in many of the water bodies, deterioration of water quality due to runoff from overburdened dumps and wastewater from tailing ponds, water scarcity, dust problems on public roads due to transportation of iron ore by trucks, poor air quality in mining areas and certain high traffic density road corridors and such other related problems. Research studies have pointed out that mining activities have a considerable negative impact on the ecosystem of Goa.

Climate and Rainfall

The entire coastal belt is characterized by plentiful, regular and seasonal rainfall during the monsoon season, oppressive weather in the hot months and high humidity throughout the year. The diurnal and seasonal variations in temperature are not large. The summer from March to May is a period of increasing temperatures. May is the hottest month of the year. In the hot months, sea breeze is enjoyable nearly throughout the day particularly in the coastal areas. Further inland, in the tract at the foot of the Western Ghats both days and nights are oppressive in summer. In North and South Kannada, it is less oppressive in summer, thanks to the forests. The onset of the south-west monsoons brings down the temperatures slightly. The south-west monsoon bursts over the region by the first week of June and the rains continue with little intermission till about the beginning of October. During the monsoons the skies are heavily clouded. The rainfall increases rapidly from the coast towards the Sahyadris on the eastern border. In the coastal strip the annual rainfall decreases from south to north, e.g. in Raigad district the average rainfall in the season is around 3000 mm, whereas in South Kannada district it is around 3900 mm. Similarly, rainfall near the coast is around 2100 mm, in Raigad district and 3000 mm, in South Kannada district. Rainfall in the Ghats increases considerably. Amboli in Ratnagiri district for instance receives 7446 mm, and Karkal in South Kannada district 5800 mm. Nearly 95 per cent of the annual rainfall is received during the south-west monsoon season. Winds are very strong during the monsoon season and blow from the south-west. In association with cyclonic storms in the Arabian sea in the post monsoon months and to a lesser extent in May, the region experiences very strong winds and heavy widespread rains. Thunderstorms appear in April and May and from late September to middle of November. With the withdrawal of the south-west monsoons day temperatures increase slightly in October and November. From October to December the winds are generally moderate and blow from the north-west. Night temperatures are the lowest in January.

Distribution of Forests along the Western Coast

Forest is a valuable natural resource of our country. They provide timber, fuel, fodder, some food, and income to rural people as well as environmental services and watershed protection. Forests form a distinct ecosystem. In India the Eastern Himalayas and the Western Ghats constitute two of the 24 global hotspots of biodiversity. Forests of Western Ghats are a source of plant and animal biodiversity. Forests are situated mainly on the higher slopes and spurs of the Western Ghats, while the lower slopes and flatter tops of the hills are cultivated. In the State of Forest Report (SFR) 2001 published by the Ministry of Environment and Forests, all coastal districts of Maharashtra and Karnataka are classed as hilly. Dakshin Kannada is also classed as tribal district. The latest statistics of area under forest in our study region as given in the State of Forest Report, 2001 reveal that half (50.4%) of the region's geographical area is under forest cover. The forests are not evenly distributed along the coastal region. Some areas have quite considerable forest resources while others have limited resources. The natural distribution of forests is controlled by terrain and climate. The construction of Konkan railway involved diversion of forest land.

Coastal Maharashtra

According to Directorate of Agriculture, area under forests in Raigad district in 1994-95 is 1680 km^2 (24.46% of geographical area). Reserve forests comprise 1275 km^2 or 82 per cent of this forest area. As per report of MRSAC and NRSA, the area under forest is less—1601.43 km^2, -26.4 per cent of it is classed as evergreen, 32.8 per cent deciduous, 30.1 per cent degraded and 4940 ha or 3.1 per cent mangrove forest. As per 1991 District Census Handbook, forest is recorded in 1162 out of 1851 villages of Raigad district Forest constitutes 21.45 per cent of district area.756.25 ha. area is under Social Forestry. Forest Survey of India's 2001 data reveals much more forest coverage (2288 km^2, 32%). Over half of it is dense forest while 47 per cent is open. Forest cover is good on the Sahyadri hills of the district, being mainly reserved forests. Maximum forest area falls under Roha tahsil—14.34 per cent of the total forest area. The reserve forests between Roha and Kolad consist mainly of teak and cashew trees. Small pockets of reserved forests occur in Murud and Shrivardhan tahsils. There is limited forest cover along the coast and in Kal-Savitri valley.

In Ratnagiri district a very small area considering the size of the district is under forest (0.82%). According to 1997-98 report of the Directorate of Agriculture, the forest cover is limited to over 6700 hectares. Half of it is found in Khed tahsil mainly on the hills. This constitutes only 3.6 per cent of tahsil area. In Lanja, Dapoli and Rajapur tahsils, 0.6 to 1.34 per cent of area is under forest, while rest of the district is totally devoid of forests. As per MRSAC report only 4300 ha (0.52 per cent of district area) is under forest cover. According to NRSA report 105 km^2 area of Ratnagiri district is under forest—87.5 per cent of this forest is classed as evergreen, 10.7 per cent as deciduous and 1.78 per cent mangrove. According to 1991 District Census Handbook forest is recorded only in 39 villages of this district 1634 ha area is under social forestry. The 1999 SFR presents the combined statistics for Ratnagiri and Sindhudurg districts. It was classified as a "hill district" as approximately 85 per cent of its land is hilly. The area under forest is 5672 km^2 comprising 43.45 per cent of district area. Over half of it (58.55%) is open forest and 41.23 per cent is dense forest. SFR 2001 presents separate statistics for Ratnagiri and Sindhudurg districts. According to this source, 26.94 per cent of the area in Ratnagiri district is under forest. 86 per cent of it is classed as dense forest. The hilly area in this district is rocky and devoid of soil cover, and hence there is hardly any growth of vegetation. The Resource Atlas of Ratnagiri district (MRSAC) point out that a wide belt running north south is almost devoid of vegetation. A travel in this region by road confirms the presence of barren rocky hills without any tree growth.

Hence the statistics presented by FSI appear to have overestimated the forest cover in this district.

Historical records show that around the year 1756 Ratnagiri district was covered with forests. The ravines and water courses were covered with fine teak and Bankot supplied the ribs and framework for the ships of Indian navy which were constructed in the shipbuilding yards at Malwan and Vijaydurg. After the transfer of the district to the British in 1818 and import of labour for cultivation, the rich forests were cleared for cultivation. In 1829, on the suggestion of the Collector Mr. Dunlop, the forests were placed at the disposal of the people. The locals considered this as a charter for unlicenced and unlimited woodcutting and sold the timber and firewood that they chopped down. Private forests were gradually denuded and to this day there is very little forest growth.

Sindhudurg District

The 1981 District Census Handbook of undivided Ratnagiri and Sindhudurg district records forest cover over 466.78 km^2. According to Directorate of Agriculture, the area under forest is only 597 km^2 or 10.51 per cent of its geographical area. Reserved forests are limited to over 17 km^2 and unclassed forests over 383 km^2. Forest is recorded in 316 villages of this district. Area under social forestry is 4057 ha, which is the largest among the 3 coastal districts. As per MRSAC 1227 km^2 or 24.12 per cent of district area is under forest. This report mentions that there is good vegetation cover in the central and eastern part. According to NRSA statistics 1159 km^2 or 23 per cent of district area is covered with forest, out of which 42 per cent is evergreen, 54.2 per cent deciduous and 3.3 per cent degraded. Forest plantation is observed over 472 ha. FSI 2001 report publishes for the first time forest statistics separately for Sindhudurg district. It reveals 45.67 per cent of district area under forest, most of it is dense. The forest is concentrated in tahsils of Kudal, Kankavli and Malwan (15 to 19 % of tahsil area) and to some extent in Sawantwadi (9.38%). The existing forests chiefly yield firewood. Very little timber is produced.

Goa

According to Directorate of Agriculture, 1255 km^2, i.e. 34.75 per cent of the state's area is covered with forest in the years 1999-2000, out of which 1055 km^2 is government owned and 200 km^2 is private forest. According to Conservator of Forests1424.38 km^2 of area is under forest in the same year—1224.38 km^2 of it is state owned or Government forest. 13.5 per cent of the total forests have been earmarked as Reserve forests and 26.8 per cent as unclassed forest. According to NRSA 2019.55 km^2 area in Goa is under forest cover—39 per cent of it evergreen, 59 per cent deciduous and 1 per cent degraded. Forest plantation is observed over 6471 ha. and forest blanks over 2460 ha. Statistics of Department of Science, Technology and Environment based on satellite imagery reveal that 1824 km^2 or half (50.26%) of the state's area is covered with forest. Goa's forest is mainly deciduous (80.6%) Evergreen forest comprises one-fifth of the total forest. As per SFR 2001, 2095 km^2 (56.59%) is under forest, a large portion (85%). of it is dense forest, and the rest is classed as open forest. Goa's forest is mainly deciduous (80.6%). Evergreen forest comprises one fifth of the total forest and Mangrove forest comprise less than one per cent (0.93%).

About three-fourth (72%) of Goa's forest area is concentrated in South Goa district and only 28 per cent falls in North Goa district. Most government forests are located in the eastern half of the state, on the slopes of the Western Ghats. The private forests are in the coastal or central part. The coastal tahsils of Salcete, Mormugao, and Bardez are almost devoid of forests. Sanguem tahsil not only has the maximum (63.7%) area under forest at tahsil level in Goa, but 45 per cent of Goa's total forest area is also concentrated in this tahsil. Rich natural forests and dense forests, with crown density above 40 per cent cover the ghats having slope above 30 per cent Patches of open forests with crown density of 10 to 40 per cent can be seen at the border of the dense forests. The midland region has few selected patches of open forests and planted forests. Teak, eucalyptus and bamboos are mainly grown here. The dominant vegetation type in Goa is moist deciduous with 1512 species of plants. The vegetation pattern in Goa shows a spectrum of variability from west to east, the gradient being correlated with ascending contours from sea coast to the crest line of the ghats. The forests of Goa also provide shelter to 48 genera of mammals, 275 genera of birds and 60 genera of reptiles.

Coastal Karnataka

Due to heavy rainfall the undulating and ravinous lands of coastal Karnataka are generally covered with thick evergreen forests with extensive forest areas. They are deciduous in nature, lofty and dense. They are rich in various species of wood such as sandalwood, rosewood and teak. Firewood and valuable timber of very good quality is abundantly available. There are plantations of quick growing trees like eucalyptus and teak. In Uttar Kannada district, as per Directorate of Agriculture, 8152 km^2 or 79.55 per cent of the district is covered with forest. As per NRSA data the extent of forest cover is more, 8202.32 ha. Over three-fourth of it (78.53%) is evergreen, 12.32 per cent moist deciduous and 7.61 per cent degraded. There is no mangrove forest as per NRSA data. As per FSI, the extent of forest cover is somewhat less–7807 km^2. It is remarkable that 83.3 per cent of it is dense. About 40 to 50 per cent of the total area of Karwar-Honavar coast is under forest. Over 75 per cent of the area of Yellapur Dandeli region is covered under thick bamboo forest. Over 90 per cent of the area of Yellapur tahsil on the Sahyadri hill slopes is under forest.

The extent of forest cover in Dakshina Kannada district is limited compared to Uttar Kannada district. According to Directorate of Agriculture 2275 km^2, i.e. 27.29 per cent of the district area is covered with forests. According to NRSA the forest cover is slightly less-2214.74 km^2. Almost the entire forest (94.61%) is evergreen and the remaining (4.29%) deciduous. Degraded forest is confined to only 0.56 per cent of total forest. FSI statistics reveal much larger area (4717 km^2 or 55.8%) under forest. Three-fourth of it is classed as dense forest. This discrepancy of data though not as large as in Ratnagiri district, is rather difficult to account for. It is most likely that the hilly areas are considered synonymous with forest land in the interpretation of satellite imageries. The per cent of area under forests varies at tahsil level ranging from 40 per cent in Coondapur tahsil to only 3.4 per cent in Mangalore tahsil. Timber of black wood, mango, tirwa and wild jack is obtained from these forests. Besides, valuable minor products such as cardamom, pepper, cashewnut, cinnamon, ginger, myrabolons, soapnut and rampatri are also found.

In Dakshina Kannada district dense forests are found covering the hills near Coondapur and are mainly of deciduous type. The interior coastal talukas have mostly semi-evergreen and evergreen type of forests. The talukas near the sea coast have secondary moist deciduous type of forests. These forests occur along the river valleys of Kali, Gangavali, Aghanasini, and Sharavati. Wild beasts such as bison, sambar, spotted deer, barking deer, leapord, elephants and porcupines are found in these forests.

Mangrove Forest

Mangroves are salt tolerant forest ecosystems found mainly in tropical and subtropical inter-tidal regions of the world. Typically they produce tangled masses of arching roots that are exposed during low tides. The ecosystem is highly fragile subjected to salinity effect, aeration and onslaught of wave action. Mangrove forests are considered to be one of the economically important, productive and bio diverse wetlands on earth. These forests are excellent breeding grounds for a variety of fish and molluscs. They occur mostly in sheltered bays protected against heavy wave action and winds. Healthy mangrove forests are key to a healthy marine ecology. For the first time in FSI 2001 assessment, mangrove cover has been classified into dense and open: Dense mangroves having more than 40 per cent crown density and open mangroves having crown density between 10 to 40 per cent As per SFR 2001, mangrove forests in our study region constitute 1.1 per cent of India's mangrove forest.

Mangrove forests have the largest extent in Raigad district (34 km). As per NRSA the extent is much more compared to SFR. These forest are found along the coast in the northern part. of Raigad district. A small extent (9 km) of mangrove forest is observed in Ratnagiri district and only 1 km in Sindhudurg district. The area along the creeks are covered by clay deposited by tidal action. These clay pockets form the mudflats and mangrove vegetation. Only 5 km of mangrove forest in coastal Maharashtra is of open type, while rest of the forest is classed as dense. The area statistics of mangrove forests in Goa vary considerably. According to Department of Science, Technology and Environment of Goa Government, mangrove forest is observed over 17 km^2. (over 0.93% of area). According to NRSA mangrove forest is observed over 4.49 km^2 in Goa. As per SFR 2001 the extent of mangrove is limited to 5 km^2 of area, 4 km^2 in North Goa and 1 km^2 in South Goa district. The mangrove forest is noticed in coastal talukas of Goa in the Chapora region, near Mandovi river and around Maem, Durgat, Panaji and Akshi. Around Chorao, a large island at the mouth of the river Mandovi the area consists of naturally regenerated mangrove vegetation and 14 different species of mangroves are found here. A mangrove nursery was established for the first time in Goa at Chorao in 1985. As per NRSA, there is no mangrove forest in coastal Karnataka. As per SFR 2001 mangrove forest is noticed in Dakshina Kannada district over 2 km of area. Small patches of mangrove vegetation are found near Coondapur and Barkur. In SFR 1999 mangrove forest was found over 11 km^2 in the north around Karwar coast in Uttar Kannada district near Mashen and Piplibag area. But no mangrove vegetation is recorded in this area in SFR 2001.

The Region as a Transport Corridor

The term corridor clearly suggests connection and access. It implies physical and linear geographical form and homogeneity. A corridor is defined as a combination of one or more important infrastructure axes with heavy flows of traffic that link important urban areas. Corridors are bundles of infrastructure that link two or more urban areas. These are highways, rail links, (high speed trains, intercity lines, local trains, trams), separate bus lanes, cycle paths, canals, sea connections and air connections.

The corridor can be seen as a dynamic space. A corridor enables free and easy flow or transmission of people, goods or information. In general corridor development concerns connections that use different transport modes and carry both passenger and freight transport. Corridors can exist as axes of infrastructure, economic development and urbanization or of institutional development. Wide ranging requirements are being served in the same corridor. The development of corridors over the course of time reflects technological advances in transport modes, and the construction legacy of different varieties of infrastructure. This is the crux of Whebell's theory of corridors. The western coastal region can aptly be called as a transport corridor. Because of the geographical location of this region, sandwitched between the Western Ghats to the east and the Arabian sea to the west, all the modes of transportation are passing through this region in north south direction. Konkan railway, National

Highway and coastal shipping route are almost parallel to each other. This corridor serves as an urbanization and economic development axis. In this corridor the new generation of infrastructure (i.e. Konkan railway) is located in the vicinity of older systems (i.e. National Highway No. 17). Economic development and urbanization are concentrated along these lines. Here the infrastructure development has occurred within the framework of a corridor which has transport chains over its full length.

Transport Infrastructure of the Region

The transport infrastructure along the western coast was not satisfactory and inadequate prior to construction of Konkan railway. Transport and communications were not well developed in Konkan's forested hilly region due to various constraints. The cris-crossing rivers and streams, rugged mountains, creeks all create a variety of obstacles with the result that the average cost of road and railway construction in the region is relatively higher than in the plains. The winding nature of roads and steep gradients in large parts of the region result in escalating operational costs. The demand for a fully developed highway materialized in the First Five Year Plan when the Bombay to Goa road or west coast road was developed to the standard of Highway (N.H. 17). This National Highway was the forerunner to the development of railway. Prior to construction of national highway, there was practically no all weather transport infrastructure in this belt. The national highway was fully saturated on many stretches. Increased traffic was progressively damaging the structurally weak roads, rendering them unusable in many tracts. Other roads were very narrow and below standards at several places. After every monsoon, the road surface were badly damaged and the process of road repairing was endlessly delayed, causing considerable hardship to the road users. Existing systems were due for replacement as free heavy cargo movement was not possible. The regional road system was plagued by the problem of missing links, especially those that connect into the deep interior. Bridges were required on some rivers. Inland waterways which are tidal in nature run perpendicular to the coastal strip and not along it. Coastal shipping could not offer round the year service due to heavy rains and rough sea conditions during the monsoon season. Important nodal ports along the western coast needed connectivity of a broad gauge railway line. Besides, as tourism grew and people from backward region felt the need to travel to larger cities, they wanted a quicker and cheaper method of getting there. The construction of a broad gauge railway line along the western coast was therefore necessary to provide a reliable and efficient transport infrastructure.

The hierarchy of road network in our country consists of (1) National Highways, (2) State Highways, (3) Major District Roads, (4) Other district roads, (5) Village roads. The transport system of the West Coast region incorporates a highway system, a coastal system, railway system, inland waterways and airways. The major transport lines cutting through this region are the road and railway connections between the Konkan and Maharashtra plateau that pass through the Sahyadri ghats. The role of road transport in the regional framework of the Western Coast is important, not only because it can open up the inaccessible areas hidden in the rugged mountainous regions, but also because it can perform the traffic feeding function for both railways and coastal shipping. Several villages in the interior are still not connected by all weather roads.

The Major transport links passing through our study region are as follows:

National Highways

The National Highways are black topped and surfaced throughout NH. 17—Panvel-Mahad-Panaji-Karwar-Mangalore-Trichur is the second longest in length in Maharashtra. Further south, it passes through all the towns of coastal Goa and Karnataka.

- NH 17-A—Cortalim-Mormugao.
- NH 17-BA—New byepass 4 lane highway connecting Mormugao port to NH 17 at Verna by skirting Vasco town is recently completed.
- NH 4A—Panaji-Fonda-Anmod-Belgaon.
- NH 48—Mangalore-Hassan-Bangalore.

State Highways

The state highways connect the coastal region with the interior. They are:

SH 40—Mahad-Bhor-Poladpur, SH 41—Panvel-Uran, SH 42—Alibag-Pen-Khopoli, SH 43—Kholapur-Pali-Ratansai, SH 44—Shriwardhan-Lonare, SH 45—Poladpur-Wai-Sirpur, SH 46—Dabhol-

Khed, SH 47—Guhagar-Jath-Bijapur, SH 48—Sawarde-Tausal, SH 49—Mirya-Ratnagiri-Miraj-Bijapur, SH 50—Musakaji-Rajapur, SH 53—Deogad-Phonda-Nipani, SH 54—Vijaydurg-Redi, SH 55—Malwan-Kasal, SH 56—Vengurla-Belgaum, SH 57—Banda-Domamarg, Panaji-Phonda-Dharwar, Karwar-Yellapur-Hubli, Ankola-Sirsi, Honavar-Siddhapur-Sangar-Shimoga, Mangalore-Beltangidi-Chikamagalur.

The National Highway connecting Mumbai with Kanyakumari was the forerunner to the development of railway and it has connected a large number of places increasing the accessibility. State highways, district roads and other lines of transport have also played a complimentary role since 1980.

Rail Lines

Prior to construction of Konkan railway, the West Coast was connected by the following rail links:

1. Mormugao-Madgaon-Londa-Hubli/Miraj, Hubli to Dandeli
2. Mangalore-Arsikere-Bangalore, Mangalore to Hassan,
3. Mangalore-Shoranur-Ernakulam,
4. Panvel-Uran (Broad Gauge). The first three were earlier Metre Gauge lines which are recently converted into broad gauge and integrated with the national railway network.

Waterways

Goa is linked with Bombay by a daily steamer/ship service during fair weather and this is a suitable mode specially for tourists, as it is cheaper, safer and comfortable. The service operates between Bombay and Panaji. Since the rivers of Goa are navigable, an integrated rail-cum-road-cum-water transport system could be devised.

The inland waterways are one of the important mode of transport in Goa. They are used for tourist boat cruises and for the transport of mineral ores. Small and large country crafts and mine ore barges are operated through the following rivers: Tiracole (27 km), Chapora (32 km), Mandovi (66 km), and its Naroa branch (10 km), Kandepar branch (26 km), Sinquerim (8 km), Mapusa (27 km), Corjuem (8 km), Bicholim (15 km), Sanquelim (18 km), Zuari (southern branch—78 km), Cumbarjua canal (18 km), Sal river (28 km), Talpona (5 km), Galgibag (2 km). Over 69 lakh passengers were ferried by inland waterways in Goa in the year 1999-2000. The northern parts of Raigad district are connected to Bombay by daily ferry services. The region also has the facilities of sea waterways.

Airways

The airports on the Western Coast are located at Mumbai, Goa (Dabolim) and Mangalore. There is also an airstrip at Ratnagiri. Domestic as well as international flights operate from Mumbai and Goa. Goa being an important tourist destination domestic flights operate daily from Goa to Mumbai, Pune and Bangalore. From Delhi flights operate to Goa for six days a week, from Agatti three days a week, from Chennai, Cochin and Kozhikode two days a week. International flights operate from Goa to Kuwait and Goa to Sharjah two days a week each. Mangalore is connected to Mumbai by daily flights and to Bangalore and Chennai by flights three days a week. A new international airport will soon be constructed in North Goa to boost up tourism. A proposal to construct a new airport at Panvel (Navi Mumbai) is under consideration.

REFERENCES

CIIs National Conference on Transport Infrastructure Emerging Scenario, (December 1997). New Delhi.

Economic Survey, (2002- 2003). Government of India, Ministry of Finance and Company Affairs, New Delhi.

Golias, J.C. (June, 2002). Analysis of traffic corridor impacts from the introduction of the New Athens Metro system, *Journal of Transport Geography*, Vol. 10, No. 2, pp. 91-97.

Ranade, P.S. (2000). Industries and Environment, a Study of Impact Assessment, APH Publishing Corporation, New Delhi.

The India Infrastructure Report (1996). Policy imperatives for growth and welfare. Expert group on the commercialization of infrastructure projects. Published on behalf of Ministry of Finance, Government of India, by Dr. Rakesh Mohan, NCAER, New Delhi.

Vaidya, B.C. (ed.) (2003). *Geography of Transport Development in India*, Concept Publishing Co., New Delhi.

2

Passenger and Goods Transportation on Konkan Railway

Profile of the Project

The jurisdiction of Konkan railway starts from the south of Roha under Central Railway in Raigad district of Maharashtra state. From here it runs southwards in Ratnagiri and Sindhudurg districts, then enters Goa (north and south Goa districts) and southwards passes through Uttar (North) Kannada and Dakshina (South) Kannada districts of Karnataka and terminates at Thokur near Mangalore under Southern railway. Out of the total route length of 760 km, Maharashtra state has 382 km (50%), Goa has 105 km (14%) and Karnataka has 273 km (36%). There are 63 stations on the Konkan railway line to provide facility of goods and passenger transportation. Frontline technologies have been incorporated in the construction of this line to make it the most modern railway line in the country today. The whole line has been designed and constructed for a speed potential of 160 km per hour. The Konkan railway alignment literally carved out of the Western Ghats passes through one of the most difficult and formidable terrain in India both physically and geologically with nearly 1500 streams to be crossed. Laying of track across rugged terrain criss-crossing rivers and plunging valleys was a gigantic task. Most of the length was difficult to approach even though some sections were close to NH 17. Main reason why such a railway line could not be constructed earlier, was the enormous cost involved in bridging various rivers and valleys and boring tunnels through several hilly ranges. The railway line has as many as 179 major bridges with a total linear waterway of 21.5 km and 1819 minor bridges with a total waterway of 5.73 km. The longest bridge on Konkan railway (KR) is 2.065 km long across river Sharavati near Honnavar in Karnataka state. In the construction of bridges a number of technological innovations have been incorporated. There are a large number of viaducts, deep cuttings and high embankments. There are 92 tunnels with a total length of 84.6 km. The bridges and tunnels are constructed at a cost of about Rs. 2000 crores. Nine tunnels are longer than 2 km each. The longest tunnel is at Karbude near Ratnagiri which is 6.5 km in length. The engineers faced their most difficult challenge in tunnelling work. Keeping in view the possibilities of air pollution inside long tunnels due to diesel traction, forced ventilation has been provided in 5 long tunnels on KR route. Jet fan technology is used in 4 major tunnels and a centrifugal fan ventilation is provided in the Karbude tunnel. This ensures desirable environment inside the tunnels. Safety measures have also been taken to protect the life of passengers in the event of accidents inside tunnels. Fire extinguishers have been installed and sufficient lighting arrangement made in all tunnels. In the construction of bridges and tunnels a number of technological innovations have been incorporated. Its massive railway structures have become admired features. Panvel viaduct in Ratnagiri district is the tallest viaduct in Asia and has piers as high as 64 metres, taller than the Qutub Minar. Despite the formidable condition of the terrain through which the alignment passes, minimum radius of curvature of 1250 metres and a ruling grade of 1 in 150 compensated have been adopted in order to facilitate smooth movement of heavy-haul freight trains. A four-aspect colour light signalling system has been adopted for the first time. KR route is without any joints or fish plates and track ballasts are not used anywhere on the track. KR route is constructed to face earthquake at 6+ Ritcher scale. Due to its engineering marvels Konkan railway itself has become a tourists attraction. The pioneering spirit and the courage to experiment with new ideas which characterized the construction phase has resulted in a very modern infrastructure.

Konkan Railway: A Trend Setter

When the KR project was conceived, it was conceived as a highly viable project with an internal rate of return of 13.5 per cent Konkan railway has been described as the greatest event since Lord

Parashuram created this region. Konkan railway is hailed as a gift to the nation in the 50th (golden jubilee) year of our Independence. Historically the responsibility of providing transport infrastructure has been vested with the government. With the formation of the Konkan Railway Corporation Ltd. (KRCL) a new era of infrastructure planning and execution through private participation dawned in the country. Konkan railway project is a trend setter in the country in many ways specially in regard to innovative approach for financing. This is the first major infrastructure project in the country taken up on "Build, operate, transfer" (BOT) concept. It is the longest railway line ever constructed in one stretch in the country. If a project of this magnitude and dimension were to be completed through budgetary grants by the Indian railways, it would have taken 25 to 30 years. A bold and unique financing scheme was therefore evolved. The BOT concept envisages that the KRCL will construct the line and operate it for a period of 10 years after commissioning. Konkan Railway Corporation Ltd. (KRCL) a separate statutory body was formed on 19th January 1990 under the Indian Companies Act, 1956 pursuant to the agreement between Government of India acting through the Ministry of Railways and the Governments of Maharashtra, Goa, Karnataka and Kerala as a joint venture with headquarters at CBD Belapur, Navi Mumbai. The beneficiary states became the partners in KRCL contributing 49 per cent of equity capital: Maharashtra—22 per cent, Goa—6 per cent Karnataka—15 per cent and Kerala—6 per cent. The Ministry of Railways contributed 51 per cent. Of the total estimated fund requirement of Rs. 3350 crores (gross) inclusive of Rs. 950 crores of financing costs, Rs. 800 crores has been contributed by these five partners as equity and the balance has been raised by market borrowings mainly through tax free bonds and external commercial borrowings to avoid time over run and cost over run. The common man, i.e. consumers invested in Bonds issued by KRCL to raise funds for its construction. Though KR does not pass through Kerala, it shared part of the expenditure because it shares the benefits of KR arising out of better links with the coastal region and speedy transport. Historical and political factors have played their part in implementation of this project.

New Work Culture

Konkan railway project will go down in history as a project well conceived and well executed through a very novel and very imaginative funding route. KRCL has invested heavily in the latest technology in areas of operation, communications and computerized working systems. Thus the requirements of manpower for KRCL are much less than those of any equal length on the older system of Indian Railways. The responsibility of repaying the staggering public debt makes it imperative that the manpower requirements are kept to the minimum workable level so as to minimize the recurring annual expenditure. As in the case of construction stage, during the operation stage as well, an entirely new work culture has been brought about by proper training and operation of the staff. The maintenance schedule and maintenance philosophy has also been completely recast, e.g. the permanent way maintenance is entirely machine oriented, eliminating the concept of regular gangs, push trollying, etc. Mobile maintenance gangs with rail mounted fully equipped vans attend to non-programmed maintenance, while two high speed tamping machines and one points and crossings tamping machine attend to programme maintenance. The result is that the requirement of permanent way maintenance staff is less than one person per track km (0.75). The concept of mobile squads has been introduced for signalling, telecommunication and electrical installations. The carriage and wagon maintenance is also based on a new philosophy, with the result that the total manpower requirement for KR for operation and maintenance is only 4.2 men per km for all departments put together as against the yardstick of 28-30 persons per km on Indian railways. World Bank has appreciated KR for its cost effective optimum utilization of resources and efficient management practices benchmarking it as a Model railway. KR has received special certificate of appreciation for excellence in cost reduction. This award is instituted to encourage the Indian industry to adopt cost reduction measures to face the challenge of globalization of Indian economy. KR a lean organization has one-third the number of employees in Indian Railways manning the same stretch of 760 track km KR has achieved excellence in cost reduction through extensive use of latest technologies and innovative design concepts and information technology systems (Ministry of Railways, PIB Press Release, 10 March, 2004). KRCL has demonstrated that even a state sector organization, with dynamic management, extensive autonomy and latest technology can perform exceedingly well, and may even excel over the private sector (Sreedharan, 1998).

The project was opened up in phases from either ends

progressively. Accordingly, Mangalore-Udupi sections in the south and Roha-Veer in the north were completed first. The dates of opening of various sections for passenger traffic are as follows:

Section	*Length in km*	*Date of opening*
Mangalore-Udupi	67	20-03-93
Roha-Veer	48	27-09-93
Udupi-Kundapura	33	05-04 95
Veer-Khed	52	25-09-95
Khed-Chiplun	30	23-02-96
Chiplun-Ratnagiri	77	16-11-96
Ratnagiri-Sawantwadi	161	20-01-97
Kundapura-Canacona	181	01-06-97
Canacona-Madgaon	32	05-08-97
Pernem-Madgaon	57	24-08-97
Sawantwadi-Pernem	22	26-01-98

Passenger Train Services along Konkan Railway Route

Konkan Railway is ensuring the environmental improvement of this region by properly channelising the tourist traffic to tourist destinations through its passenger friendly services. Prior to introduction of KR, there was only one train to perform journey along the western coast of India from Trivandrum (Kerala) to Mangalore (Karnataka), i.e. Parashuram Express. It had a special appeal because of the day time journey it used to make along the coast. During 1995-96 two pairs of passenger trains were running on the Roha, Chiplun section, whereas one pair of passenger train was running on the Mangalore-Kundapura section. Passenger train operations were extended from Chiplun to Ratnagiri on 16-11-1996. One pair of passenger train plying between Dadar and Chiplun was extended to Ratnagiri. This was further extended from Ratnagiri to Sawantwadi on 20-1-1997. 468 km out of 760 km were commissioned for goods and passenger operations by 31-3-1997. One pair of Express train was introduced between Kurla and Sawantwadi. The entire stretch of Konkan railway was declared open for through traffic between Roha and Mangalore w. e. f. 26 January, 1998. However, the first passenger train was diverted via Konkan railway from 21st March, 1998. Eight passenger trains per day started operating on this route in July 1998. In an effort to make the region more accessible to tourists some trains originating from Delhi, Rajkot and Gandhidham and going towards the south, including the Delhi-Trivandurm Rajdhani Express were diverted on the picturesque Konkan railway route. Express and passenger trains running over Konkan railway route include the prestigious 2431-2432 Nizamuddin-Trivandrum Rajdhani Express (twice a week), 6335-6336 Gandhidham-Nagarcoil Express (once a week), 6333-6334 Rajkot-Trivandrum Express (once a week), 6337-6338 Hapa Cochin Express (once a week), Daily trains are: 2617-2618 Mangala-Lakshadweep-Hazrat Nizamuddin Ernakulam Express, 2619-2620 Kurla (Lokmanya Tilak Terminal LTT)-Mangalore Matsyagandha Express, 6635-6636 Kurla-Trivandrum Netravati Express, 0111-0112 CST (Chhatrapati Shivaji Terminal)-Mumbai Madgaon Konkan Kanya (overnight), 0103-0104 CST-Madgaon Mandovi Express (Daytime). Passenger trains are: Dadar-Ratnagiri, Diva-Sawantwadi, Sawantwadi-Madgaon, Mangalore-Madgaon and Karwar-Madgaon (DMU). A new high speed Jan-Shatabdi Express was introduced between Mumbai (LTT) and Madgao (Goa) from the summer of 2002. As announced in Railway Budget 2003, a new weekly Trivandrum-Jodhpur Express train via KR route started from 1-11-2003. Goa (Madgaon junction) can be accessed by all the trains running on Konkan railway. However, Mandovi (daytime), Konkan Kanya (overnight) and Jan Shatabdi Express are available exclusively for travelling from Mumbai to Goa and back. Special "Jan Chair Car" coaches of Jan-Shatabdi actually second class look like AC coaches because of the construction and give the traveller a true 21st century travel experience. The tourist traffic on this route has picked up fast. All the day time and night time trains run to full capacity. Konkan railway is planning to increase the speed of trains running over its tracks from the present 120 km per hour to 160 km per hour. In order to bring the commercial capital Mumbai closer to the tourist destination Goa, the trial run of the country's fastest train was conducted at a speed of 150 kmph. The distance of 442 kms between Roha and Madgaon was covered in a record time of three and a half hours (KR Press Release, 25-2-2003). After a morning tea in Mumbai, a passenger is expected to reach Goa before lunch and the travel time between Mumbai and Goa would be further reduced or another superfast train in the future could be introduced.

Passengers Travelled by Various Trains

In the Survey conducted by RITES in May 1994 regarding passenger traffic, it was anticipated that on completion, KR will be able to move over 6 million passengers per year. In order to study seasonal fluctuations and trends of passengers travelling on Konkan railway route, monthwise statistics of bookings for trains done at Mumbai Passenger Reservation System (PRS) from January 1999 to September 2001, i.e. for a period of 33 months are utilized. Number of passengers travelled and left waitlisted for these trains are as follows:

Table 2.1: Passengers Travelled by Trains on KR Route

(Average number of passengers travelled per month)

Year Train No.	0103	0104	0111	0112	2618	2619
1999	6167	5410	18859	11002	1657	14413
2000	13967	11799	21673	14651	2572	18509
2001*	17480	13788	26631	17289	2311	23552
Total number of passengers travelled						
1999	74006	62528	226308	99019	19891	172961
2000	167605	141586	260083	175816	30862	222105
2001*	157325	124100	239680	155603	20798	211967
Passengers waitlisted (Total)						
Total of all trains						
1999	21990	69666				
2000	22027	112209				
2001*	20208	104338				

*Upto September

Source: Konkan Railway Corporation, Corporate Office, Belapur.

The above statistics reveal that the total number of passengers travelled per year and passengers travelling per month is on increase. This increase also reveals the growing popularity of KR. The public patronage for these trains is very encouraging. This is also reflected in the statistics of waitlisted passengers. The total number of waitlisted passengers by all the trains has increased. Their monthly average increased from 5806 in 1999 to 11,593 in 2001. Every year the maximum passengers travel in the month of May, followed by December and this number has also increased since 1999. The lowest number of passengers is recorded in the month of July, being the rainiest month. Among all the trains given in the table, the most popular train used by the largest number of passengers is the Konkan Kanya Overnight Express from Mumbai to Madgaon. Seasonal or monthly fluctuations by this train are the least meaning thereby that this train always runs to its full capacity from the very beginning. In

January 1999, passengers travelling by Konkan Kanya constituted 51 per cent of total passengers of all trains considered above, whereas they constituted 37 per cent of passengers in September 2001. Their absolute number has more than doubled. Number of passengers left waitlisted is also the highest for this train. The monthly average of passengers on the waiting list increased from 1833 in 1999 to 2245 in 2001. The largest number of waitlisted passengers are found in the month of April-May followed by November-December. The share of waitlisted passengers of this train to total waitlisted passengers for all trains which was 32 per cent in 1999 has come down to 19 per cent in 2001. The main reason for the popularity of Konkan Kanya Express is its convenient timings. This train leaves CST, Mumbai station at 10.30 p.m. at night and reaches Madgaon at 10.25 a.m. next day after 12 hours of journey. By overnight journey, a tourist saves a precious day of his holiday. The corresponding train from Madgaon to Mumbai leaves Madgaon at 6.15 p.m. and reaches Mumbai at 6.35 a.m. Though the number of passengers travelling by this train is lower than its counterpart, it is also on increase. A daytime train from Mumbai to Madgaon (0103 Mandovi Express) is not so popular as the Konkan-Kanya Express. The number of passengers travelling by this train, though less, is also on increase. The number of passengers travelling from Mumbai to Goa is more than those travelling from Madgaon to Mumbai. The largest number of passengers travelled in the month of November, and the lowest in the month of July. Train No. 2619 from Mumbai to Mangalore Matsyagandha Express ranks next in popularity to Konkan-Kanya Express as revealed by the number of passengers travelled. This is the only train which covers the entire stretch of Konkan railway. It leaves Kurla (Lokmanya Tilak Terminal) at 3.15 p.m. and reaches Mangalore at 7.30 am next day. Mangala Express (2618) from Hazrat Nizamuddin to Ernakulam is a long distance train diverted on Konkan railway route. This has substantially reduced the travel time of passengers going to Kerala and hence attracting increasing number of passengers, more during October to December compared to other months.

Madgao Station Gets ISO Certification

The Madgao railway station has become one of the few railway stations in the country to achieve the prestigious ISO 9001 : 2000 certification for providing better amenities and services to ticket holders. In fact it is the second A class railway station in the country after Secunderabad to have attained this unique distinction. A Norwegian firm was entrusted with the work of certification of the quality management system of Madgaon station. Facilities that come under the ambit of the certification and provided at Madgaon station for ticket holding passengers include UV treated drinking water, sanitized toilets, refreshments, waiting halls, effective illumination, parking area, dissemination of information, first aid facility cloak room, luggage carting, tourist taxi counter, general stores, STD booths, book stalls, cyber café, milk parlours, juice stalls among other facilities. This station building is designed by famous architect Charles Correia and has a spacious concourse for passengers. Keeping in view the heavy rainfall in the region the roof is sloping. The station has a main platform with shelter and island platform both 560 mts long to accommodate 24 coaches. Platforms where express trains halt are at a high level and are provided with concrete slabs and interlocking paver blocks (*Nav Hind Times*, Goa, 17 Januery 2006).

Along with the number of passengers, earnings of Konkan railway from passenger traffic have increased considerably. As per 11th Annual Report of KRCL, its total traffic earnings increased from Rs. 80 crores in 1998-99 to Rs. 151.65 crores in 2000-2001. The Corporation has done very well in 2005 and is striving to better its performance continuously (*Nav Hind Times*, Goa, 17 January 2006).

Special Trains/Holiday Specials

In order to cope up with the extra rush, to clear the waiting list of passengers and to cater to their pressing demands and holiday mood Konkan railway runs special trains during summer and winter/ Chrismas vacations and during special festivals like Ganapati, Dassera, Diwali and when a large number of people visit their homes in Konkan. Holiday special trains during day time enable tourists to enjoy the line sideview. To clear extra rush of passengers during various festivals and Yatras lika Bharadi Devi, KR runs unreserved special trains between certain stations, additional temporary halts are provided to certain trains and extra coaches are attached for the convenience of the passengers. Following a heavy demand by the passengers from the coastal Karnataka, the timings of certain trains have been revised. Two special unreserved trains were run daily, one from LT Terminal to Madgaon (Goa) and another from CST to Madgaon and back from 17-8-2001 to 2-9-2001. Two express and one passenger train were augmented. Konkan railway introduced the following special trains for summer season during April, May and June 2002 : CST Mumbai-Madgaon—CST Mumbai Daily Total 96 services, Mumbai-Mangalore Weekly—Total 22 services, Mumbai Ernakulam Weekly- Total 22 services, Vasco-Howrah—Vasco—Total 6 services, via Madgaon, Ratnagiri, Kalyan, Nagpur, Bilaspur. During the Ganesh festival in the month of September 2002 Konkan railway in cooperation with Central Railway introduced 56 special trains for the passengers from 6th September to 22nd September. Out of them 28 trains were reserved trains. 0105/0106 LTT-Karwar and back reserved trains had 20 coaches each. 0107 and 0108 CST- Madgaon and back unreserved special train had 14 coaches. Apart from these, Madgaon-Sawantwadi-Diva passenger train and Dadar Ratnagiri passenger trains were provided with 4 extra general coaches from 6th September to 23rd September. Special arrangements were also made for passengers boarding these special reserved trains from Thane and Panvel stations. Two second class sleeper coaches were kept reserved for Thane passengers and 3 sleeper coaches for Panvel passengers. For reserved trains, separate coaches were also earmarked for passengers from Thane, Diva and Panvel for unreserved special trains. In 0106 up special train to Mumbai one second class sleeper coach each was earmarked for passengers from Sawantwadi, Kudal and Kankavli for return journey. Taking into account the increase in speed of trains over Konkan railway and introduction of the new Jan Shatabdi Express extra halts are provided for certain trains. According to the press release of March 2004, 204 special trains would be run on the KR route between Dadar Ernakulam, Dadar Mangalore and CST Madgaon to clear the summer rush during April to June. The trains will have 10 second class sleeper coaches, one two-tier and one three-tier AC coach, two general coaches and two guard-cum-brake vans.

A special train runs from New Delhi to Ratnagiri and back in the month of December every year (during 21st to 31st December) to cater to special tourist traffic of followers of Nirmala Devi, a well known spiritual personality, who holds special camps and performs mass marriages at Ganpatipule, a beach resort near Ratnagiri. A large number of her overseas followers also participate in this ritual. The train has 3 AC coaches, 20 sleeper coaches and a few unreserved coaches. This train has proved to be so popular that it has a long waiting list of passengers. The largest number of special services introduced to Goa reveals the popularity of Goa as a holiday destination and increasing demand of tourists for train travel. All the newly introduced trains have proved to be extremely popular. They have attracted sizeable passengers and tourists traffic. These trains have enhanced the carrying capacity of tourists and are a boon to tourist industry. The future business plan for Konkan railway include an urgent expansion of line capacity along with improvisation. The proposed business plan would include running of more trains following an improvement in line capacity. However, the plan will neither include doubling of tracks nor any electrification. KRCL in its original master plan was to have 82 stations. After the business proposal comes into effect, the number of stations will be increased, as also the trains (*Nav Hind Times*, Goa, 17 January 2006).

Railways and Goods Transportation

Railways have traditionally been the bulk carrier of various commodities. Railway has promoted enormously the internal and external trade of the country. Railways are best suited for long haul transportation and are particularly well suited for carrying dry bulk commodities in cases where water transportation is not an alternative. The railways' role in carrying manufactured commodities is limited by high terminal cost, large capacity and low speed. The line haul costs of rail transport are lower than those of trucking, but rail terminal costs are much higher. For long hauls, railways can become competitive in terms of speed. Indian railways are the most important hauler of domestic freight traffic in the country. Though the share of railways in goods transport has declined steadily for the past more than two decades, almost two-third of rail earnings are derived from freight traffic. Railways are more suitable and economical for carriage of large volumes of any commodity offered in train loads irrespective of distance. Indian Railways have identified need of a new segment of transportation, i.e. parcel section. Parcel express trains are introduced on several routes connecting metro cities. Increased movement of freight traffic is caused by all-round development in different sectors of the economy. The increasing industrialization has led to the transport of more and more oil, liquefied gas, petrochemicals and other goods. The volumes carried as a single cargo have also increased. A study of "West coast transport network" sponsored by Government of Maharashtra (Modak, 1980) while emphasizing the necessity of construction of Konkan railway had noted that "a large amount of freight traffic is originating in Bombay and the northern part of the country for being moved to southern part of the West Coast and *vice versa*. It does not find road transport economical for long haul. Coastal shipping has its problems of accessibility and seasonality. The only mode which satisfies the long haul need in respect of a majority of commodities is rail transport."

Goods Transportation: Initial Problems, Anticipated Traffic and Current Scenario

Commodities for transportation originate in areas of production and terminate in areas of consumption, final demand or point of trans-shipment. A survey conducted at 16 major check posts in South India in 1996 (Indian Railways, June 1996) revealed that "Mumbai is the destination for goods from Kerala and Tamil Nadu. From northern region of Kerala coconut, copra, coconut oil are transported to Mumbai and Ahmedabad. On an average 48 trucks a day carry 428 tonnes of goods in this direction. From Cochin spices, tea, coffee move to Mumbai for export. An average of 88 trucks transport 840 tonnes of rubber and its biproducts. From Coimbatore and Erode region about 42 trucks carry 426 tonnes of cotton and textiles. Vanaspati is moved in bulk from Gujarat to Ernakulam and Calicut. All this traffic can be easily attracted by trains. The railways have not tapped the potential and need to undertake massive marketing to gain business from the existing customers."

Compared to passenger traffic, goods traffic on KR presents a bleak scenario. The freight traffic on KR line is not as high as expected. Indian railways had envisaged at least 8 goods trains to run, of which 5 trains to run from north India to south India along this route because of the tremendous advantage in the haulage distance. Goods trains were not allowed to run in the initial stage. If one goods train is run on this line, there will be Rs. 10 lakh net saving, net benefit to the Corporation. If 8 trains are run, that means Rs. 80 lakhs per day earning which is not available to the Corporation. This is the reason why financially the project is facing a serious problem (Sreedharan, 2001). It was anticipated in West Coast Transport Network Study (1980) that the various projects and developmental programmes in the future may generate substantial new traffic. The traffic survey done by the Southern Railway was later updated by M/s RITES in1995 and 1999. Industrial traffic and traffic diverted from road were calculated separately in the report. It was anticipated that on completion, KR will be able to move 7.47 million tonnes of freight during 1998-99 and 7.64 million tonnes during 1999-2000. However, the actual statistics reveal that 1.26 and 1.61 million tonnes of freight was transported during these years which is 17 to 21 per cent of the projected traffic.

Initially the Indian Railway sections adjacent to KR, viz. Shoranur Mangalore on Southern railway and Diva- Roha on Central Railway were not equipped to handle additional traffic expected to materialize on the commissioning of KR. There was inadequate line capacity due to single line in these adjacent divisions. There was mismatch of freight train loads resulting from load restrictions owing to severe gradients in Mangalore-Thokur and Kulem-Madgaon sections. There were inadequate loop capacities and inadequate terminal/transfer facilities at Mangalore and Roha. The constraints in the smooth transfer of traffic between Konkan railway and the adjoining sections of Indian Railway were gradually removed. Diva-Panvel and Diva-Vasai doubling work was completed by 1999. Three crossing stations on Panvel-Roha section were commissioned in December 1996. Appropriate procedures were established for dealing with load mismatch between inward and outward goods trains at Roha and Thokur. Terminal/transfer facilities were created at Roha, Mangalore and Thokur.

Earnings from freight traffic to the Corporation are given below in Table 2.2:

Table 2.2: Freight Traffic Earnings of Konkan Railway

(*Rs. crores*)

Year	*Earnings, Projections*	*Actual*
1998-1999	234.35	16.67
1999-2000	240.70	20.44
2000-2001	N.A.	32.25
2001-2002	N.A.	34.26

Source: 10th and 12th Annual Reports of Konkan Railway Corporation Ltd., 1999-2000 and 2001-2002.

The experience in the operational phase since 1997 points out that the financial and commercial projections made in the original traffic survey have been way off the mark. The earnings of the Corporation have not been as per original projections made in the Survey. The comparative figures in the Table reveal that freight traffic earnings were only 7 to 8 per cent of the projected earnings.

Among the major industries planned are a super thermal power station at Nandipur Mulki, and Madhabad near Karwar. The government of Goa is also planning a thermal power station. In Maharashtra, the Enron's gas based thermal power station near Dabhol in Ratnagiri district became operational. A pig iron plant has been set up at Redi in Sindhudurg district. A steel plant is also coming up at Satarda in Sindhudurg district. Each programme is expected to generate both freight and passenger traffic.

The 1980 Transport survey gives a long list of large and medium industries under execution, proposed and visualized in the west coast region. The present industrial scenario reveals that many of the proposed and visualized industries have not come up. Original traffic pattern on which the projections were based, has changed dramatically during the seven-year construction period, due to changes in the economic scenario of the country. The goods traffic pattern which emerged from the initial survey for this project have undergone tremendous changes. Three major factors which were not envisaged at the time of original traffic survey relate to:

(a) Import of food grains at southern ports considerably reduces the movement of food grains from north to south.

(b) Development of coastal shipping between coastal Gujarat to Madgaon/Mangalore. Ever since the construction of Konkan railway project began, coastal shipping has developed tremendously, and a lot of traffic which was anticipated to move over this line has been diverted through low cost shipping alternative.

(c) Due to economic recession, specially slump in steel market, a number of investment plants in the Konkan region, like Usha Ispat and Nagarjuna Steel have either been cancelled or postponed. This has severely affected the availability of traffic to be carried over Konkan railway line.

The industrial traffic identified by RITES is in respect of the movement of raw materials and finished goods for a number of existing industrial units along the new railway line as well as new units, either planned or in process of being set up. The principal units in this category are shown below in Table 2.3.

Table 2.3 Commodities for Movement

Particulars of unit	*Commodities for movement*	
	Inward	*Outward*
Usha Ispat, Redi, Dist. Sindhudurg	Iron ore, limestone	Steel products
Vashisti Detergents Ltd., Lote, Dist. Ratnagiri	Oils, minerals	—
Zuari Agro Chemicals, Goa	—	Fertilizers
Narmada Cement, Ratnagiri	Clinker, Gypsum	Cement
Rashtriya Chemicals and Fertilizers, Raigad dist.	—	Fertilizers
Sesa Industries, Goa	—	Steel, pig iron
Hindustan Petroleum Corporation, Mangalore Refinery	—	POL & LPG

Source: Salelkar, S.V. (Ed.) *A Treatise on Konkan Railway*, Konkan Railway Corporation Ltd., Navi Mumbai, 1999. pp. 142-43.

RITES had identified 21 industries in respect of the movement of raw materials and finished goods. Among them, only the industries listed in table have actually come up and utilize the Konkan railway transport for the inward or outward movement of their raw materials or finished products. Rest of the industries have not yet come up due to certain constraints or industrial recession. Those which have come up have made their own arrangements of transportation. Even the industries listed send only part of their products through Konkan railway, e.g. in case of Mangalore refinery, only 3 to 4 rakes are handled by Konkan railway per day. Raw material for Vashisti Detergents, Lote (liquid chemicals) comes by road and the outward movement is also piecemeal. In case of Narmada Cement, raw material comes by sea route from Gujarat and dispatch is done partly by Konkan railway and also by road. About 100 trucks are dispatched per day to Kolhapur/Satara. Fertilizers manufactured at Rashtriya Chemicals have local market and do not use railway. Raw material for Finolex industries, Ratnagiri, i.e. natural gas is imported by sea route, and finished products (resin) 600 m tonnes per day is sent mainly to Pune by road. They do not use railway. In the wake of subsequent developments, the traffic anticipated in the case of M/s Sterlite Industries have been dropped due to local problems, and M/s Supreme Petrochemicals, Nagothane have indicated that KR would not be relevant for movement of their products. The food grain traffic from Punjab to Kerala has not materialized through KR and the HR steel traffic also did not materialise, causing a drop in revenue. For the year 1998-99 the KRCL got the major streams of traffic of food grains, HR Steel Coils and POL from MRPL. New flow of traffic generated during subsequent years includes fertilizers and iron ore, sponge iron, cement and petroleum products, clinkers, LPG, etc. Since the economy is recovering, it is expected that the goods traffic on KR should develop in due course.

Goods/Freight Transportation on Konkan Railway Route

The Konkan railway has made it possible for finished goods and raw materials to be transported both quickly and cheaply. Transporting goods from Vadodara to Mangalore meant taking them across 2390 km. On KR this distance is cut down to 1283 km, i.e. a saving of 1107 km. From Kota to Cochin products like marble, fertilizers and cement had to be taken across 2735 km. On the Konkan railway this has been reduced to 2240 km, i.e. a saving of 495 km and a saving of 17 hrs. Such enormous savings in distance and time has brought tremendous benefits to the industry. With the availability of this reliable and faster transport, it is possible for the perishable sea food products and fruits in refrigerated wagons, to have access to new much more lucrative big city markets.

Material required for construction of Konkan railway, mainly cement, was brought by goods trains. During construction phase the godowns of cement, etc. were at Roha. Goods trains passing over Konkan railway route carry the commodities originating from Northern Indian stations and go up to various South Indian destinations. Goods traffic from Ratnagiri station consists mainly of rice, wheat and fertilizers. From June to December 1998 Ratnagiri handled 275 wagons of rice, 504 wagons of wheat and 53 wagons of fertilizers. In the year 1999, 613 wagons of rice, 672 wagons of wheat and 228 wagons of fertilizers were handled. In February 1999, wagons of wheat originating from Gujarat, Punjab and Haryana had destinations in Kerala, Ratnagiri and Goa. Wagons of rice originating from stations of Punjab, Haryana and Jammu Tawi were sent to destinations of Vasco (Goa) and Ratnagiri, wagons of food grains from Punjab were sent to Goa. Fertilizers originating from Trombay were sent to destinations of Mangalore, Hubli and Shimoga town.

For food grains the terminating traffic is booked in North India to the following stations on Konkan Railway: Ratnagiri, Verna, Karwar and Udupi. Cross traffic is booked from North India to Mysore, Davangere, Vasco. In case of fertilizers, traffic originating from Verna is mainly booked to stations in South India. Traffic booked from Thal, Trombay, Bharuch on the Western coast is terminated at Ratnagiri and Verna. Sizeable cross traffic is also booked from Thal. Trombay, Panamburu to stations of South India through Konkan railway. In case of cement traffic booked at Wadi is terminated to Verna. Cross traffic of cement from Wadi to Mangalore is booked on concessions offered by KR. These concessions were subsequently discontinued. In case of iron ore, cross traffic is booked from Ranjitpura/Vyasankers to Panamburu. In case of LPG, originating traffic is booked from HPCT siding, Thokur to Cheralapalli (S.C. Railway) via Madgao. POL originating traffic is booked at HPCT siding, Thokur to Navalur/Desur on South-central railway via Madgao and to other stations on Southern railway. Pig iron cross traffic was introduced in 2001 and was booked to stations in North India via KR on offering station to station rates.

A summary of goods traffic which moved through Konkan railway route is given below:

Freight Summary: (No. of trains)

	1998-99	*1999-2000*	*2000-01*
1. Total terminating trains	57	91	102
2. Total originating trains	35	67	110
3. Through/cross traffic	97	153	157

Facilities for Goods Transportation

Separate platforms and goods sheds are constructed at stipulated stations of Konkan railway for handling goods traffic. Ratnagiri (Maharashtra), Verna (Goa) and Udupi (Karnataka) are the major freight loading and unloading stations on Konkan railway. These stations have special facilities for handling goods traffic, i.e. loading and unloading of rakes, storage yards, goods sheds, and they can handle one full rake of 40 wagons. Kudal and Chiplun are the intermediate freight stations which can handle half rake. Ratnagiri has emerged as a centre of goods traffic after the establishment of new industries at and around Ratnagiri. The indigenous manufacture of 4000 hp new generation micro processor controlled diesel locomotive used by Konkan railway is the only freight locomotive on Indian Railways capable of hauling 58 BOXN wagons on 1 : 150 gradient, as prevailing on KR route. It can run without maintenance for 3 months.

Konkan Railway's Schemes of Goods Transportation

Konkan Railway faces the challenge of business development and marketing for diverting traffic from vessels and the road. In order to attract traffic from roadways and improve its market share, the Konkan railway had initially planned to introduce two freight trains originating from Mumbai and Ahmedabad and going up to Kochi. It was envisaged that the train would accept peacemeal cargo from traders all along the route as opposed to the current practice of accepting a minimum cargo to fill a wagon. The proposal aimed to provide a multipoint consolidation and distribution of cargo with the help of private agencies. It had guaranteed on time and in tact transportation to its customers. The decision was an attempt to boost goods traffic on the KR line. The services were supposed to start in April 1999. The KRC reportedly shelved its plan of starting two freight trains. This followed the poor response received from trade circles (*Monthly Review of Indian Economy*, July 1999, CMIE). In order to run the KR on business principles and in pursuit of furthering its commercial interests, KR is exploring several innovative schemes. The Corporation is making all efforts to develop new streams of traffic to compensate for the loss of traffic, which was originally expected to come. KR may also think of starting parcel express trains in future.

"Hire Your Own Wagon" Scheme

To encourage short distance, general commodity freight movement, Konkan railway has introduced a "Hire Your Own Wagon" scheme. Salient features of the scheme are:

(1) Full wagon from any station to any station on Konkan railway (Surathkal to Kolad).

(2) The capacity of each wagon is approx. 58 tonnes.
(3) Loading, unloading, sealing and locking the doors will be the responsibility of the Party.

Tariff: From Mangalore end to Roha end—Rs. 39,250 per wagon

From Roha end to Mangalore end—Rs. 26,990 per wagon (The rates applicable for any station to any station on Konkan railway).

Konkan railway has recognized the need to be customer friendly to attract high rated traffic. For freight/parcel business the agents have their own PCs and modem to connect and through passwords gain access to the system to register demands and obtain allocation information. The schedule of the movement of train by which the booking will be cleared is also available to the agent, so that he can advise the customer and match his loading operations.

Container Trains

"Container service" has been introduced between Kolad and Verna. KR is trying to establish contacts with Container Corporation of India to attract container traffic. KR is associating with the project of expansion of Cochin port, which is being developed as international Container Terminal. KR will be its associated partner in transporting the containers towards north. The potential of transporting 5 to 6 trains per day is identified. It is also exploring the possibility of transporting iron ore of Hospet for export through Hubli-Ankola section of railway line (12th Annual Report, 2001-02, KRCL).

Roll on Roll off Service

The Konkan railway project is a trend setter in the country in regard to the innovative approach in various sectors, particularly the introduction of "Roll-on Roll-off" service. On the background of tremendous scope for movement of goods by rail on the western coast of India, Konkan railway's ro ro service is the most appropriate and timely decision. By means of piggy back operations and through the establishment of "park and ride" facilities at railway stations—rail and road transport complement rather than compete with each other. As the Konkan railway is alert and attentive to the needs and demands of trade and industry, it introduced this innovative goods transportation system for the first time in India on its newly commissioned railway infrastructure. It started on 26th January 1999, year after the inauguration of the first direct passenger train service. Ro ro is a "truck on train" concept or a "piggy back service". Piggy back literally means on shoulder or back of another person. Here loaded road vehicles (trucks) are carried on top of railway freight vehicles (flats/ wagons). The piggy back service has already proved popular in Britain and European countries. The Gotthard tunnel in Europe provides piggy back transport of 4.2 m high lorries under the Alps and across Switzerland by rail. The suggestion for extending piggy back services on Indian railways was put forth in the Report of the National Transport Policy Committee of Government of India, Planning Commission, as early as in May 1980, but it had not been implemented on any section of Indian railway so far. This service combines the advantages of door to door service by road transport with economy in cost for long distance traffic provided by railways. This makes it possible to achieve the objective of consolidating multi point generation of low volume traffic into train loads and multi point distribution. Konkan railway is playing the role of an associate partner in road transport. KRCL brought out an illustrative brochure in English and regional languages containing information regarding this service in order to give it a wide publicity. It also contains the Ro ro route map. Konkan Railway's website also gives some details of this service. The following analysis is based on the data collected by this researcher during personal visits to the office of KRCL at Belapur, Navi Mumbai from time to time.

Ro Ro's Operating System

Under this truck on train concept, loaded goods trucks are rolled on the railway flats at Kolad, 145 km south of Mumbai and are carried "piggy back" till Surathkal, 20 km from Mangalore. At this place they are rolled off. Hence this service is known as Roll on roll off (ro ro). If the trucks are going north, they roll on at Surathkal and roll off at Kolad. The jurisdiction of Konkan railway starts after Roha station. Roha is the point of handover from Central Railway to Konkan Railway. Kolad nearest to Roha is the first station on Konkan Railway route. It is also located on Mumbai to Goa National Highway 17. Surathkal, 20 km from Mangalore is the southern most station under KR's jurisdiction. Availability of required parking space for trucks at Kolad was an added advantage for the introduction of this service from here. A platform and a fixed ramp are constructed at Kolad railway station. The ramp starts from outside the station near a siding so that the trucks can be driven on to train, i.e. railway flats. Portable ramp is used for unloading. Acid tankers are loaded and unloaded from portable ramp. Similar facilities are constructed at Surathkal for movement towards Mumbai. These are scheduled daily services operative both ways. Daily departure time of ro ro train from Kolad is 1700 hrs and from Surathkal it is 1500 hrs. This time schedule is fixed taking into consideration the convenience of the transport operators to load the trucks and assemble at the station for rolling on the flats. One rake of flats runs every alternate day as it takes about 24 hours to travel the distance of 734 km from one way. The rake returns after being loaded. A driver and a cleaner are allowed to travel free on board the truck and also enjoy the train ride and good sleep without any tension or worry. They drive the trucks to their destinations after rolling off at Kolad or Surathkal railway stations. The delivery is guaranteed within 12 hours. The trucks for ro ro are booked through the private transport companies having their offices in Mumbai, Calicut and Mangalore. Vehicles can be booked at Kolad railway station and Verna railway station one day in advance. Konkan railway's information technology aided tracking system keeps track of trucks every minute of the way.

KRCL operates the railway line with the rolling stock leased from Indian Railways. This occasionally puts certain limitations on its operations. The diesel locomotive for the ro ro train is taken on lease from Central Railway. The flats or open wagons of ro ro train are taken on lease from Ministry of Defence. Each train has 30 flats (wagons) to accommodate 60 ten tonner to 14 tonner trucks. Each flat carries two trucks. Permissible dimension of truck height is 3.4 m from road level.

The Advantages of Ro Ro Service

This service helps to partially put an end to the tiresome and hazardous truck journey from Mumbai to Mangalore on the Western coast route by saving 734 km of stretch of travel without any tension. Due to very heavy seasonal rainfall and the nature of terrain in the coastal region, road journey is more hazardous and time consuming specially during the monsoons. Hence the monthly statistics reveal that this service is more popular among the operators during the monsoon season. This service allows faster transportation of goods due to faster turnaround of trucks. Transporters can use the same fleet of trucks for more number of trips leading to faster turn round and higher profits.

By regular road route, it takes 3 to 4 days for a single trip. Ro ro service ensures delivery within 24 hours both ways. The service available between Kolad and Verna (Goa) saves a 430 km stretch of travel for Goa by National Highway on Ghat section. Thus both consigners and consignees are benefited by quicker transit time. Truck owners are benefited by better turn round of trucks in Mumbai-Mangalore-Mumbai circuit due to saving of fuel, manpower and wear and tear of trucks. The service ensures on time and in tact delivery free of any risk factor and without multiple handling of consignment. At national level this service helps to conserve precious diesel, decongest national highways and reduce the pollution level along the route. Ro ro service is a major positive step towards environmental improvement. It assumes significance on the background of increase in environmental pollution associated with transport sector. The ro ro is eco friendly and the ecology as well as economy are benefited by it.

Customer satisfaction is far more important in case of a new service. If a customer once attracted to ro ro cannot be accommodated in it, he turns to another mode of transportation and refuses to come back. It is difficult to attract him again. Customers have a distinct preference for road transportation in view of: (1) cheaper road freight, (2) quick transit time, (3) door to door delivery offered by road transport agencies. When they realized that ro ro service can also offer these advantages, this service became immensely popular among the truck operators.

Commodities Transported

The exact details or break up of goods loaded in each truck are not recorded with transport companies or with KRCL. The quantum of any of these commodities is also not available. They are grouped into general commodities, parcel, textiles, medicines, rubber and spices. The commodities transported by trucks in ro ro service are sponge iron, marble, chemical powder, zinc powder, stone powder, aluminium powder, animal food, MS steel plate, A.C. sheet, food grain, rice, lime powder, grocery, biscuits, machinery. POL products (petroleum, oil, lubricants) are not transported by ro ro. There is no transportation of mangoes, the major produce of Konkan coast by ro ro trains. This is because there is no intermediate booking or loading station for ro ro in Ratnagiri or South Konkan, the major mango producing region and therefore, the mango exporters are not availing this service.

Sulphuric Acid

Konkan railway has succeeded in attracting this commodity for transporting in the ro ro. Sulphuric acid tanker transportation was introduced in ro ro service on 22nd February, 1999. Sulphuric acid is used as raw material in the manufacture of phosphatic fertilizer. The manufacturing units of Sulphuric acid are located at Vashi and Navi Mumbai. From these units sulphuric acid is sent to Zuarinagar (Goa) phosphatic fertilizer plant. Special permission is taken from Chief Controller of Explosives, Nagpur for transporting sulphuric acid tankers by ro ro. Loaded tankers are rolled on at Kolad and roll of at Madgaon. It takes 10 hours for the rake of acid tankers to reach their destination in Goa. The tankers are empty in their return journey from Madgaon to Kolad. Madgaon is simply the midway loading and unloading station. Freight service does not originate from here for either direction. Such service is not yet introduced due to non-availability of additional loco (diesel engines). Transportation of sulphuric acid stopped since January, 2000 due to some technical problems.

Transportation of trucks and tankers on ro ro during the initial years is given below in Tables 2.4 and 2.5.

Table 2.4: Tankers Loaded on ro ro Service

Month/year	*2 axle*	*3 axle*	*Total*
July 25 to 31,1999	26	3	29
August 1999	108	21	129
September 1999	158	25	183
October 1999	217	24	241
November 1999	240	34	274
December 1 to 15, 1999	97	11	108

Table 2.5: Transportation of Trucks by ro ro Service

Month/year	*Kolad to Surathkal*		*Surathkal to Kolad*	
	No. of trucks	*Rakes*	*No. of trucks*	*Rakes*
January to December 1999	2741	168	1607	145
January to May 2000	2277	64	—	—

Source: Data collectedd from KRCL's office during personal visits.

Initially there was some resistance from transport operators, truck drivers and from those who thought their interests might be adversely affected by this service, like wayside *Dhaba* (restaurant) owners. However, due to effective marketing strategy, Konkan railway could identify potential customers and persuaded them to use this innovative service. To begin with, ro ro service was operating thrice a week or every alternate day due to limited availability of flats. When additional rakes became available from Defence Ministry, ro ro operated daily from both the sides since March 1999. However, due to Kargil war, Defence Ministry rolled back 15 flats for defence transportation. Therefore, from May to August this service again operated every alternate day. In September 1999, 60 rakes were newly acquired from Defence Ministry for expansion of this service and one rake of 30 DBKM (flats) operates daily since then.

The monthwise statistics of movement of trucks and rakes reveal the initial problems the service had to face because Konkan railway itself was in its nascent stage. Konkan railway also faced difficulties in operating the service due to non-availability of loco. The loco of ro ro was attached to passenger trains in case of their engine failures, resulting in delays of ro ro. Locos are sent to to Kalwa or Parel locosheds of Central railway for repairs. The rake movement was the lowest in the month of June. The position of rake and truck movement could not pick up in July as well. The officials also admitted that the monsoon had been a difficult period for them. The service had to be closed for 8 days. The position gradually improved thereafter. There was an increase in demand and very good response to this service from the transporters as revealed in the sudden spurt in statistics since August. Truck owners had to wait for 24 to 48 hours at the terminal to be rolled on the flats. The trend of increase in traffic has continued since then. But what is remarkable is the one way (downward) traffic of ro ro on Kolad Surathkal route. The upward (Surathkal to Kolad) traffic which exceeded the downward traffic in the initial months declined thereafter and totally stopped since January 2000. Rakes are not going up to Surathkal since February 2000. They terminate at Verna in Goa. Formerly Madgaon was the unloading point. The tankers are booked only up to Madgaon and do not go further south to Surathkal. There is no loading point of ro ro at Ratnagiri, as it is short distance and hence not economical to operate to railways or traders.

Freight Rates

The rail and road comparison shows that road trucking is cheaper than rail for short hauls in respect of most commodities The advantage turns in rail's favour in medium and long hauls in respect of most of commodities. Ro ro service from Kolad to Madgaon (350 km) illustrates medium haul and Kolad to Surathkal (734 km) long haul transportation. The freight charged by Konkan railway for this service is moderate and easily affordable by small operators, which has contributed to its popularity. The service has proved to be popular and financially viable. The truck owners admit that it works out to be very cheap for them, and are enthusiastic about this service. With reasonable tariff rates, ro ro is fulfilling the expectations of transport operators who handle the long term traffic over the west coast route. At present there is daily ro ro service operating both ways between Kolad and Verna. The freight charge increases as per the axle of the truck.

Table 2.6: Freight Rates of ro ro Service

(Rates in Rs.)

Section	*Trucks 2 Axle*	*Trucks 3 Axle*	*Tankers*	
	*gross weight up to 25 MT**	*gross weight up to 35 MT**	*2 Axle*	*3 Axle*
Kolad Verna	5100	7650	5500	8250
Verna Kolad	4800	7200	2750 For empty	4125 return

* For every additional tonne or part thereof the freight rates increase by Rs. 200. Weight bill + tare weight produced and declared by sender for trucks is considered for charging.

With the transportation of over 2700 trucks during the first year of its operation (January to December 1999) Konkan railway's ro ro service earned revenue of over Rs. 1.37 crores. During the period of January to April 2000 it earned an average monthly revenue of Rs. 19.32 lakhs and a total of Rs. 77.3 lakhs. Even though the freight tonnage and earnings through freight traffic in subsequent years have increased compared to previous year, the Corporation has to go a long way to improve its own business activity and to mobilize as much traffic as possible to improve the performance. Indian Railways transported a record all time high revenue earning freight during

2003-04. This indicates buoyancy in the core sector of the national economy. (PIB Press Release, 12 March 2004, Ministry of Railways).

Roll on Roll of Service

The automobile policy of the Government has given an unprecedented impetus to the growth of personalized mode of transport in India leading to high energy consumption and traffic congestion. The past couple of decades have witnessed an increase in vehicular traffic of all kinds of vehicles on our highways. This is accompanied with increase in number of vehicles involved in road accidents, number of persons killed and injured. Freight transportation has increased in scale and efficiency and modern diesel powered lorries are providing dramatically increased level of community service. The traffic census of NH17 carried out by PWD of Government of Maharashtra shows that there were 5172 and 2474 goods vehicles at the two counting points Palasphata and Indapur respectively in Raigad district on 31-12-1997. The census records reduction in number of goods vehicles with increase in distance from Mumbai. There were an average of 1430 goods vehicles in Ratnagiri district and 861 in Sindhudurg district during the traffic census carried out from 19-5-1997 to 26-5-1997 (average count at 3 stations in each district). It is estimated that there is a considerable reduction in number of trucks plying on Mumbai-Goa Highway after the commissioning of Konkan Railway. Improved connections between ports and rail terminals have improved the environment of adjacent zones by taking several trucks each day off the roads.

Statistics of road accidents by type of vehicles reveal that about one-third of total road accidents are registered to be caused by goods vehicles alone. Goa ranks next to Delhi and is ahead of other metropolitan cities in terms of vehicle ownership. The fatal accidents in Goa rank among the highest in the country. Prior to commissioning of Konkan railway, i.e. in 1989-90, 2716 vehicles were involved in road accidents in Goa. In the year 1998-99, 3805 vehicles were involved in road traffic accidents wherein 225 persons were reported to be killed and 1856 injured. In the 3 coastal districts of Maharashtra 3347 vehicles were involved in road accidents in 1994-95 in which a total of 409 persons were reported to be killed and 2668 injured. Among the vehicles involved in accidents, 1170 or over one-third were goods vehicles. Buses constituted another one-fifth of total vehicles. These vehicles mainly ply on the National Highway and reveal the intensity of traffic hazards. In Uttar Kannada and Dakshina Kannada districts 488 persons were reported killed and 4213 injured in road accidents in 1993-94. The true picture of safety on Indian roads may be much worse than what the above statistics indicate because many accidents go unreported. Moreover, Indian roads have among the lowest ratio of vehicles to kilometres of roads. Accidents can never be completely prevented, they will inevitably occur from time to time especially with the great increase in transport in terms of load or of distances. Efforts are made to improve the design and operation of ships, road tankers and railway cars to reduce the number of accidents. Streets require urgent repair, resurfacing and widening to reduce traffic congestion and accidents. The diversion of goods vehicles from the Highways have a positive impact on traffic congestion and road accidents.

Advantagtes of no reservice

In our country the bulk of liquid fuel is used in the transport sector. The country is heavily dependent on imported oil. Railways provide the cheapest, energy efficient form of land transportation consuming one-sixth to one-eighth of the energy per tonne km consumed by any other mode of land transportation. It is therefore in the national interest to maximize the use of rail transportation for movement of passengers and goods. Each loaded ten tonner truck requires about 400 litres of diesel for road journey of 734 km from one end of Konkan railway (Kolad) to the other (Surathkal). When the flats of ro ro train are loaded with full capacity of 60 trucks, it leads to substantial saving of diesel-24,000 litres per trip. On this background the consumption of diesel by loco of goods train is much less. It is estimated that for 1000 km of haulage of goods train 12,000 litres of diesel is required by the engine. The consumption of diesel by the engine of ro ro train per trip comes to approximately 8800 litres. Thus it amounts to a net saving of over 15,000 litres of diesel per ro ro trip. The benefits in terms of saving of repair costs and damage are beyond measurement. The wear and tear of truck tyres during the long journey is avoided, which results in additional benefits of increasing automobile tyre life and saving maintenance cost. Moreover, the tension or worry of both truck drivers and headache of transport operators is reduced.

To conclude, Konkan railway has made it possible for goods to be transported both quickly and cheaply. The ro ro service of goods transportation has a beneficial impact on the environment of this region.

REFERENCES

Carpenter, T.G (1994). *The environmental impact of Railways*, Wiley Chichester.

Infrastructure in India: Economic Intelligence Service, CMIE, Mumbai, November 1996.

Joseph P.T. (June 1996). Critical analysis of goods traffic in South India, Railway Vs. Road, *Indian Railways,* pp. 19-20.

Modak, S.K. and D.M. Nachane (June 1980). *West coast transport network-an optimal intermodal mix*, A study sponsored by the Government of Maharashtra, Mumbai.

Railway fare and freight committee report, Vol. 1, Ministry of Railways, Government of India, December 1993.

Ray, S.K., (1995). *Tranport Planning for Developing Countries*, Prentice Hall, New Delhi.

Regional transport survey of Maharashtra state, (1966). Vol. 1, Part 1, Report. Regional Transport Survey Unit, Government of Maharashtra, Bombay.

Report of the National Transport Policy Committee, Government of India, (May 1980), Planning Commission, New Delhi.

Socio-economic Review and District Statistical Abstract for (1) Raigad, (2) Ratnagiri and (3) Sindhudurg districts, (1997-98). Bureau of Economics and Statistics, Government of Maharashtra, Mumbai.

Sreedharan, E. Ideas that have worked, A lecture at the Department of Administrative Reforms and Public Grievances on 16th January, 2001. http:/persmin.nic.in/arpg/Sreedharan.htm

3

Land Use Changes

Land use refers to man's activities and various uses carried on land. Sauer defined land use as " the use to which the entire land surface is put." Land use is also defined as "use of land by man usually with the emphasis on its prime role in economic activities." Land is a basic natural resource. Land use pattern indicates the resource base of an area. Economic development of a region depends on proper utilization of our land resources. Land use components have a tendency to show spatial and structural changes in response to variations in physical, economic and cultural factors and the process of development. The assessment of availability of land resources and changing land use pattern bring out the bottlenecks in sustainable development of land resources. It is a necessary prerequisite to know the present or existing use of land in order to plan for its optimum use. The land use within a particular region undergoes changes as a consequence of economic activities. Changes in land use reflect the impact of development activities. Changes in land use are basically governed by competitions between different types of demand. This demand is generated from urban areas, industrial and tertiary activities, from non-industrial sector and infrastructure development. With the construction of transport routes there are irreversible changes in land use or sometimes severe damage to natural resources. Each geographical area tends to have its own unique characteristics that could lead to a different pattern of land use and transport interaction.

This chapter studies the impact of construction of Konkan railway on land use of the surrounding region. The first part presents the land use along the proposed alignment and within the corridor of 20 km along the proposed alignment (10 km on both sides of railway line) at the time of acquisition, the actual land use of the land acquired for construction, and land use of selected villages whose land was acquired. The analysis of changes in land use along the railway is presented. The second part gives a detailed account of changes in forest cover in the districts and tahsils through which KR is constructed. The third part analyses the changes in other categories of land use after the completion of this project by comparing the land use at the beginning of construction of Konkan railway.

Data Sources

The statistics of land use are collected from the following sources:

(1) For the Environmental Impact Assessment (EIA) studies for Konkan railway line, a corridor survey using satellite data (IRS-LISS-II) was conducted by NRSA. The report of this survey gives the land use statistics along the proposed alignment and within the 20 km corridor.

(2) District level land use/land cover statistics were generated by NRSA using Remote Sensing Techniques under the nation-wide land use/land cover mapping project. They are based on the imageries for the year 1988-89, i.e. prior to construction of Konkan railway, and are published in the Report on area statistics of land use/land cover. Land use maps on 1 : 250,000 scale based on the above data were prepared at district level and are available for all the coastal districts of our study region. For Goa the satellite imageries of LISS 1 dated 22 October 1988 and 11 February 1989 were utilized and the land use map was compiled at Centre of Studies in Resource Engineering at IIT Bombay. For UK and DK districts, IRS LISS 1 imageries dated November 1988 and February 1989 were used and the land use maps were prepared at Karnataka State Remote Sensing Technology Utilisation Centre, Bangalore.

For Goa, land use statistics were also compiled under the project "Integrated Mission for Sustainable Development for the State of Goa", using Remote Sensing data (IRS 1 A/1 B, 1992-93, 1 C 1996) by the Department of Science, Technology and Environment, Government of Goa. Land use maps under this project are not yet published but the district state level statistics compiled by them were made available during the personal visit to this office.

(3) Wastelands Atlas of India compiled and published by NRSA gives districtwise and categorywise statistics of wastelands of India for the year 1988-89.

(4) Resource Atlas of Maharashtra State published by Maharashtra State Remote Sensing Application Centre, (MRSAC) Nagpur contains forest and other land use statistics of both Kharif and Rabi seasons based on Landsat TM imageries for 22nd October, 1988 and 11 February 1989 and April 1990.

(5) Forest Survey of India (FSI, Ministry of Environment and Forests, Government of India) publishes forest statistics based on satellite imageries at district level in the State of Forest Report published biannually. The latest statistics available from FSI are those published in the State of Forest Report, 2001, which is the documentation of the 8th Assessment of the country's forest cover pertaining to the year 2000. They are compared with the statistics published in The Report of 1993, which contain the results of 4th assessment of the country's forest cover pertaining to the period 1989-91.

(6) The Village Directory of District Census Handbooks of 1991 Census of India have published land use statistics at village and tahsil/taluka level for the year 1989-90. Comparable statistics for the 2001 Census are not yet available.

(7) District Statistical Abstracts (handbooks) published annually by the Bureau of Economics and Statistics of respective state governments contain the land use statistics including forests supplied by Directorate of Agriculture of each state. Their source is the Annual Season and Crop Report. Comparative analysis of land use changes at tahsil and district level is based on data available from these sources.

Land Acquisition for Infrastructure Projects

Infrastructure development projects, which are of crucial importance for local, regional and national development involves three processes, viz. land acquisition for the project, displacement caused by land acquisition and rehabilitation of the displaced population or payment of compensation to the affected population. There have been large scale investments both on account of internal generation of capital and increased inflow of foreign investments, thereby creating an enhanced demand for land within a shorter time span. The quest for land is on increase for export processing zones, industrial parks and housing estates. The development of infrastructure is accompanied by increased resistance and opposition to land acquisition and the resultant involuntary displacement. Infrastructure projects including construction of new transport routes, involve acquisition or redirection of use of land that are owned or utilized by individuals and communities. The most significant long term impacts of new railways result from land acquisition. Land acquisition and intrusion by railways had been a source of contention since before 1830 with the introduction of railways in Great Britain. The complexity of land acquisition is increased by the need for linear alignments of road and rail routes. This is because each of these requires 'a dedicated guide way or corridor' of its own. However, railways have the least problem with strip development and capacity expansion through land acquisition because modern signalling, rolling stock and locomotives can easily double the capacity of existing rail lines without requiring more land. Since land is our most scarce resource, it is an asset valued dearly by everyone in India. This makes land acquisition and land management for transport infrastructure projects a significant issue and it is necessary that every new road or railway line is evaluated carefully for land acquisition.

Land Acquisition for Konkan Railway

The most crucial task in the pre-construction phase of any development project is land acquisition or the consent of the owner to begin work on his land. There are constraints of land acquisition from individuals who have legal rights as well as those who are squatters. Therefore, in the initial stage the requirement of land needs to be carefully and correctly assessed. The experience in connection with land acquisition for development projects reveals that each piece of land acquired has a story of agony and a specific war fought by the affected generation. In case of Konkan railway the experience was different. An entirely new strategy was evolved to take possession of the land needed for this project. A very pragmatic and very practical approach was adopted for the construction of Konkan railway.

Proper acquisition proceedings laid down under the law were completed and land legally handed over to Konkan Railway Corporation. Land acquisition is a lengthy process. Normally the formalities for the acquisition process can take up to five years. Here the work could proceed as planned much before land was actually acquired and possession taken over. The social acceptance of a project by affected people may be critical for a project to proceed without costly delays and adjustments. In most cases of KR the land owners' consent was obtained and construction work took place unhindered. This went a long way in speeding up work on the project.

The extent of land required for Konkan railway was carefully determined on the basis of the height of the bank, depth of cutting, station development, extra land for quarries of earth, murum, stone, diversion of streams and roads. Land was also necessary for casting yards, girder assembly yards and enabling works. Though single line at present, the provision for a future double line was made on the west and land required for future doubling was acquired Land had to be acquired for tunnels as well, since the owners' rights over a land extend not just on the surface, but below the surface too. A uniform width of 20 m was acquired over tunnels.

Acquisition of Government Land

The agreement executed by the Railway Board with the four State Governments of Maharashtra, Goa, Karnataka and Kerala provided for transfer of Government land free of charge to the Konkan railway. The state governments' maritime departments control some rivers and their permission was required for alternative uses and/or future disputes on river land. Since the beneficiary states were partners in the Konkan Railway Corporation, civil authorities provided help where required, and this expedited the possession of land. There was no land acquisition for this project in the tribal territory. Only private land was acquired.

Acquisition of Private Land

Land was acquired for this project through the state governments through which the line passes. Though land related law suits are common in the Konkan, most of the land owners in Maharashtra, Karnataka and Goa willingly came forward to give consent for the acquisition of their land for this project. Village level meetings were conducted to make land owners aware of the importance of this national project and many were so convinced that they gave consent letters much before they received the compensation, so that work could start immediately. During this time, 42,000 land owners were called upon and land at a cost of Rs. 144.88 crore was negotiated. KRCL discovered that at least half a dozen people were not even aware that they had inherited the land in Konkan, since their families had neglected to do the paper work. KRCL helped them in completing the legal formalities and paid them compensation that was due. Of the 234 villages affected in Maharashtra, there were problems only in five villages. These were solved without delay by persuasion, official help and speedy compensation. The cheques of compensation were delivered to the people's doorsteps.

Care was taken to acquire minimum possible land so that ecology was affected a minimum. KRCL avoided cutting habitats wherever possible. Farmers and their fields were provided with easy communication passages along the railway line. Heritage sites were left untouched.

Current land acquisition procedures in India consist of two phases:

(1) Making cash award as per market value, and
(2) In court where a judge awards a supplemental amount.

Private land for a project of this nature is acquired under the Land Acquisition Act (LAA) of 1894. For the acquisition of land under the Land Acquisition Act, 1894, Special Land Acquisition Officers of the rank of Additional Collectors with full powers under the LAA were appointed in each district. In Maharashtra, one officer for Raigad district was headquartered at Mangaon, for Ratnagiri district at Ratnagiri and for Sindhudurg district at Kudal. In Goa there were two such officers, one at Panaji for North Goa and the other at Madgaon for South Goa district. In Karnataka there was one officer at Karwar for Uttar Kannada district and the other at Udupi for Dakshin Kannada district.

Under Section 3 of the LAA a notification was issued by the Government to enable the railway staff to enter the selected villages to ascertain the survey numbers of the land in each village to be acquired. Thereafter, the surveyors from the Land Survey department of the State Government inspected the land to be acquired and prepared a land plan, showing the land area required with survey numbers for each village. After the issue of Notification by the Government, the land required was marked with the help of village revenue maps, the staff ascertained the survey numbers of the land and area to be acquired, and details of its ownership from revenue records.

A notification under Section 4 of LAA was then issued in the Official Gazette, and also published in two local newspapers, one of them in the regional language. Almost all notices under section 4 were issued in 1991. After this the surveyors prepared the survey plans, showing the exact area of the land to be acquired. They also listed particulars of owners and tenants, the number of trees according to their varieties and structures on the property. Retired revenue staff were employed to pick up particulars of land holdings from revenue records. With all these measures and assistance to the Special Land Acquisition Officer, the work of acquiring the land was completed in record time. The Special Land Acquisition Officers worked out 80 per cent of the estimated compensation to be paid. Thereafter Konkan Railway Corporation was directed to place the compensation amount at the Officer's disposal and the Corporation issued the necessary bank drafts/cheques in the names of interested parties. A date was then fixed for taking over possession. The date was notified to all the people concerned. Eighty per cent compensation was then paid before the land was taken over by the Special Land Acquisition Officer and handed over to the Konkan railway officer. Table 3.1 furnishes the details of land acquisition.

The figures of land acquisition sectorwise reveal that more land is acquired in Karwar sector (Uttar Kannada District) compared to other sectors. This is because 760 ha. of land was acquired in this sector for compensatory afforestation. The amount of compensation received suggests that the type of land acquired in Ratnagiri sector was comparatively less productive or belonging to the category of wasteland and hence the compensation was very low. Land acquired in Udupi sector (Dakshina Kannada district) was agriculturally more productive and hence the compensation was the highest in this sector.

Table 3.1: Land and other Resources Acquired for Konkan Railway Project

Type of land acquired		*The structures/trees acquired*	
Barren land	2730 Ha	Masonary houses	6,375
Agricultural land	1761 Ha	Thatched houses	9,481
Garden land	520 Ha	Trees	201,216
Urban land	100 Ha	Crops	231 cases
Government land	150 Ha	Wells	124
Land for afforestation	760 Ha	Other structures	502
Land at Nerul (for colony)	4 Ha		
Total	6025 Ha		

Land acquired sector-wise

	Mahad	Ratnagiri North	Ratnagiri South	Kudal	Panaji	Karwar	Udupi
Land acquired	608	726	882	759	866	1555	819

Source: Salelkar, S.V. (ed.) *A Treatise on Konkan Railway,* Konkan Railway Corporation Ltd., Navi Mumbai, 1999, pp. 90-92.

Land use along the Konkan Railway Alignment Prior to Construction

In order to assess the impact of the proposed railway line on the environment of the West coast and to provide information on the broad distribution of land use pattern within a corridor of 10 km on either side of the proposed railway line, the Konkan Railway Corporation commissioned RITES to undertake the Environmental Impact Assessment (EIA) studies. A corridor survey using satellite data (IRS-LISS-II) was conducted by NRSA, Hyderabad. Its objective was to provide the base line information on the current land use/land cover pattern within the corridor. To obtain cloud free data and for better discrimination of agricultural lands from forest area, the data of post monsoon season was selected and a single date data was used. This analysis was confined to three sections: Roha-Dasgaon, (Raigad district 45 km) Panaji-Karwar (Goa,130 km) and Karwar-Udupi (Uttar Kannada and Dakshina Kannada districts 295 km). The distribution of the land use pattern along the proposed alignment and within the 20 km corridor was mapped and the areas of different land use categories were tabulated. The land use/land cover pattern along the alignment and within the corridor with related statistics as given in the Report of EIA studies is presented in the following tables.

Table 3.2: Land Cover/Use along the Proposed Alignment

Land use category	*Section*		
Area in Ha	*Roha-Dasgaon*	*Panaji-Karwar*	*Karwar-Udupi*
Forest*	12	86	532
Mangrove	—	70	20
Agricultural	208	275	700
Rivers/Estuaries	7	15	34
Salt pans	—	200	40
Built up land	—	15	—

*Total forest including scrubs identified from satellite imagery.

These estimates are based on the following assumptions:

(a) Land required at stations = 75 mts. on either side of the proposed railway line.

(b) Land required at tunnels = 10 mts on either side of the proposed railway line.

(c) Land required at all other places = 50 mts.

Table 3.3: Land Cover/use in the 20 km Corridor of the Proposed Alignment

(*Area in Ha*)

Land use category	*Section*		
Area in Ha.	*Roha-Dasgaon*	*Panaji-Karwar*	*Karwar-Udupi*
Settlements	360	142	750
Industries	245	5	0.1
Dense forest	18,175	65,663	28,895
Degraded forest	4,349	180	40
Mangroves	—	179	77
Cropland	9,087	16,573	180,093
Agricultural plantation	187	20,191	3,455
Current & other fallow	37,036	30,735	9,234
Scrubland	8,563	3,664	4,900
Upland without scrub	160	382	—
Barren rocks	682	—	1
Tanks/fishponds	545	250	52
Streams	1,500	4,392	3,200
Swamps, marsh, wetlands	—	1,454	716
Salt pans	—	885	7
Quarrying/mining	—	1.47	26
Beach sand/sandy area	—	1,347	980

Source for Tables 2 and 3: Environmental Impact Assessment Studies for West Coast (Konkan) Railway line. Corridor Survey using satellite (IRS- LISS-II) data, Report submitted to RITES. National Remote Sensing Agency, Hyderabad (1991-92).

The above tables reveal that land use pattern is more diversified within the 20 km corridor than along the alignment. Only 5 categories of land use are found along the alignment, whereas, over 15 categories of land use are enumerated within the 20 km corridor. Agriculture and forest are the two major land use categories within the corridor. The proposed alignment affects agricultural land of 1183 ha (53.4 per cent of total land area) and forest area.(including mangroves) of 720 ha. (32.5 per cent of land area along the alignment). Within the corridor of 20 km agricultural land constitutes 66.77 per cent of total land area, while forests comprise 25.54 per cent Rivers, water bodies and salt pans comprise 13.3 per cent of the land along the alignment, whereas they constitute 3.2 per cent of the land area within the 20 km corridor. Built up land is limited along the alignment only in Panaji-Karwar section while in the 20 km corridor it is found in all the sections. Land use by major categories within the 20 km corridor and the expected impact of the proposed alignment as predicted in the above report is given below.

Forest Land

Roha-Dasgaon Section

The EIA studies pointed out that the proposed alignment of Konkan Railway in this section passes mainly through scrub, agricultural land and to some extent through lower reaches of the forest and hence does not pose any threat to the environment. No mangrove formations are situated along the proposed alignment. It was observed during field visits that Reserve forest between Roha and Kolad consist mainly of teak and cashew trees. Studies by RITES indicated that the tree density of Mangaon Kolad region varies between 60 to 200 per hectare. The area has a low diversity of species. In addition, 7 ha of land under springs and small streams was affected. Forests constitute 28 per cent of total land area in this section. The degraded forest area constitutes nearly 25 per cent of the total forest area. While the upper reaches are mainly of deciduous forest, the lower slopes of hills are covered by cashew trees, scrub and grass. The isolated hills along the proposed alignment are covered with open scrub. The proposed alignment passes mainly through lower reaches of the forest and hence does not pose any threat to the environment.

Panaji-Karwar Section

The EIA studies observed that since the proposed alignment passes through secondary moist deciduous types of forests and scrubs, it does not pose any threat to the environment. No destruction of tall evergreen forests is involved. Further a series of tunnels are proposed to be constructed wherever the alignment is expected to cross the hills in the Canacona range. Moreover, it does not pose any threat to the existing flora and fauna of the corridor. The mangrove vegetation near Mashen and Piplibag falls along the proposed alignment. The extent of the mangrove fringes affected is around 80 ha. The mangrove formations of Mandovi and Chapora are not disturbed. The salt pans near Kotal with an area of around 40 ha will be affected by proposed alignment. Moreover, 60 ha of the marshy/wet land near Kotal will also be affected. To mitigate this loss the report suggests to develop appropriate stands of wetland species of plants along the embankments which besides enhancing the scenic beauty will also strengthen the bunds and prevent the erosion.

A unique feature of Goa is that a major portion of the forests in the vicinity of villages in this region consists mainly of cashew plantation. About 45 per cent of the forest area in the corridor is classed as dense forest. Its considerable extent is deciduous. Semi evergreen forests are concentrated on the eastern border of Goa and Uttar Kannada district. On the northern side there is very little forest land. The southern part of Goa between Cancona and Quepem consist mainly of dense forest. About 50 ha of forest cover is either degraded or scrub and the dense forest cover is localized in the Cape Roma-Karwar section which is 26 km in length. Along hill slopes there are a number of forest plantations. These hill slopes have been clear felled and planted with eucalyptus and teak. Most of these hills are under community control. In this section the mangrove vegetation is found in the areas adjacent to the Diwadi island and close to Karwar. The mangrove formation of Mandovi and Chapora region (over an area of 200 ha) are not disturbed by the alignment.

Karwar-Udupi Section

The coastal talukas of Honavar and Kumta have a very high grazing pressure and these areas support only degraded forests. The EIA studies observe that the proposed alignment passes mostly through scrubs and minor forests. The coastal hills and plateaus which are low in height are mostly covered with scrubs. 125 ha of dense forest is found in the part of alignment between Karwar and Ankola. The forests here are mostly of semi-evergreen type. The fish ponds are not affected by the proposed line since they are situated at a distance of about 3 to 5 km away from the railway line. The salt pans are seen extensively near Gokarna. About 70 ha. of these salt pans near Hiregutti falling in the proposed railway line may be affected. The extent of forest area affected by the proposed alignment is about 80 ha. in DK district and 350 ha in Uttar Kannada district. Dense forests constitute 12 per cent of total area of this section. The dense forests are found covering the hills near Coondapur and are mainly of deciduous type. There are no mangrove forests in the corridor of UK district. Some patches of mangrove vegetation are found near the coastal areas of Coondapur and Barkur. About 20 ha of mangrove near Coondapur may be affected due to the proposed alignment. Further, a series of tunnels are proposed to be constructed wherever the alignment is expected to pass through hills.

The EIA studies observed that no wildlife sanctuaries are located along the proposed line and the alignment does not pose any threat to wild life, flora, fauna and natural ecosystem of the area. The findings of EIA studies on the project regarding forests along the alignment were forwarded to the MOEF for their assessment and approval for the diversion of forest land.

Agricultural Land

The total agricultural land includes cropland, agricultural plantations, current and other fallow. The NRSA estimated that the agricultural land likely to be affected on account of acquisition for the proposed alignment was around 275 ha. Out of it, 208 ha was in Roha Dasgaon section. It was under paddy cultivation and mango orchards. It constitutes 57.4 per cent of total area in this corridor In Panaji-Karwar section it constitutes 46.5 per cent of total area. In Karwar-Udupi section agriculture is the main land use in the corridor constituting 83 per cent of total area. In the greater part of the area paddy is the predominant crop. In addition to paddy mango orchards are also grown extensively. Other cash crops grown are cashewnut, coconut, banana, arecanut.

Wetlands/Marshy/Salt Pans

No wetlands or salt pans are affected in Roha-Dasgaon section. The proposed railway line passes through the wet lands in the coastal areas of other sections. Near Panaji and Koral in Goa they comprise 1.6 per cent of area. The extent of wetland is limited in Karwar-Udupi section comprising 0.3 per cent of area. A number of salt pans are observed near Goa Velha and Kortal. Around 40 ha area of salt pans near Kortal are affected by the proposed alignment. The fishing activities are confined to the coastal belt and none fall on the proposed alignment. To mitigate the loss of wetlands, it was recommended to develop appropriate stands of wetland species of plants along the embankments to strengthen the bunds.

Tanks, Streams, Rivers

The proposed alignment passes through a number of rivers in all the sections. Area under streams is much more in Panaji-Karwar section than in other sections. The streams comprise 3 per cent of total area in the corridor of Panaji-Karwar section, while they comprise less than 2 per cent in other sections. In Goa most of the rivers are navigable and form vital links of rural transport. The extent of tanks is more in Roha-Dasgaon section compared to Panaji-Karwar section.

Barren Rocks/Beach Sand

As the railway alignment is in the interior areas of Roha-Dasgaon section, barren rock is the predominant land use comprising 0.8 per cent of the area. No beach sand or sandy area is observed in this section. Beach sand extends over 0.8 per cent of area in Panaji-Karwar section and over a smaller area in Karwar-Udupi section.

Mining/Quarrying Area

Presence of large scale mining/quarrying activities was noticed in Panaji-Karwar section near Ashnode. Within the 20 km corridor the mining areas occupied around 1.5 ha in Goa and 26 ha in Karwar-Udupi section. These mining areas were essentially open cast in nature. The region surrounding the mining areas consisted mainly of scrub.

Industrial Use

Due to industrial backwardness of the region, the area occupied by industries is extremely limited (0.05 per cent of total area). It is noticeable only in Roha-Dasgaon section. There is a fairly large industrial complex measuring an area of 245 ha. near Roha where the railway station is proposed. The complex is developed by

Maharashtra Industrial Development Corporation. There are three industrial units in this complex: Wymen Garden, Sethi Nippon and Powder, and Metal and Alloys Ltd. In addition there are a number of small scale chemical industries in this area. Industries are confined to only 5 ha of land in the Panaji-Karwar sector. There are cashew processing units and those manufacturing tyres and pharmaceutical products. None of the industries are situated along the proposed alignment. Land use pattern in the corridor of Karwar-Udupi section indicates the presence of only small to medium scale industries such as salt manufacturing units at Gokarna. This is in contrast to Roha-Dasgaon section. The Government of Karnataka has declared Uttar Kannada district as industrially backward, and the talukas of Udupi and Karkal have been categorized under zone III areas identified for industrial growth. On account of the reallocation of space for urban expansion brought about by the proposed alignment and other projects there is likely to be considerable impact on the environment.

Built up Land

The area occupied by built up land includes settlements in both rural and urban areas. There was rapid acquisition of lands for industrial and residential purposes by CIDCO (City and Industrial Development Corporation) in Panvel, Roha and Dasgaon talukas of Raigad district. Industrialization is expected to be accelerated by the proposed alignment. This may lead to population growth and may have local impact on the environment. In certain areas near Madgaon and Vasco there is rapid urban expansion and the proposed alignment with a railway station at Madgaon further accelerate the urban expansion which may have indirect local impact on the environment.

Land use of Land Acquired for Konkan Railway

During the construction phase, the construction work of Konkan railway was divided into seven sectors. Each sector comprised of about 110 km. It does not necessarily correspond with a district. Land use statistics of the land acquired in each sector are given in Table 3.4. It reveals that the maximum land acquired for KR (42.4% of total) is barren land followed by Agricultural and Garden land (35.4%). Forest land constitutes only 6 per cent of the total land acquired. Land use within the 20 km corridor reveals that both agricultural and forest land have a much greater extent. Barren land has the largest extent in

Ratnagiri district followed by Sindhudurg district. Both agricultural and garden land has the largest extent in Panaji sector (Goa). In addition to Goa, garden land is confined to coastal Karnataka. In Panaji and Karwar sectors. (Goa and UK districts) 28 ha of marshy land is acquired. Land for afforestation as mentioned in the Table is compensatory afforestation. Due to acquisition of forest land in Goa and coastal Karnataka, almost double the extent of land is acquired for compensatory afforestation, mostly in Karwar sector.

Table 3.4: Type of Land use of the Land Acquired for Konkan Railway

(*Figs. in Hectares*)

Land use	*Mahad*	*Ratnagiri North*	*Ratnagiri South*	*Kudal*	*Panaji*	*Karwar*	*Udupi*
Barren	265	513	733	495	218	250	256
Cultivated	316	236	98	259	304	170	318
Forest	5	—	—	—	36	272	77
Garden	—	—	—	—	192	34	48
Marshy	2	—	—	—	15	13	—
Govt. land	20	22	21	5	—	2	120
LACA+	—	—	—	—	41	780	
Other*	—	—	30	—	—	31	—

* Other—includes 4 ha of land acquired at Nerul for staff colony, 16 ha of Sea Bird project land of Indian Navy and 15 ha of land temporarily acquired in Karwar, 30 ha of urban land belonging to Maharashtra Industrial Development Corporation under industrial estate in Ratnagiri district.

Source: Salelkar, S.V. (ed.) *A Treatise on Konkan Railway*, Konkan Railway Corporation Ltd. Navi Mumbai, 1999. pp. 90-92 +LACA- Land acquired for compensatory afforestation.

Village-wise Land use of the Acquired Land

The details of land use of the entire stretch of land acquired were compiled by the respective Special Land Acquisition Officers specially appointed for this purpose by the Government in the seven sectors. The record is maintained in Basic village level registers of land acquisition at the Corporate office of Konkan Railway. The record includes area acquired and number of persons affected in each village and the details of land use, i.e. agricultural, non-agricultural and wasteland, land under water bodies, canals, roads, settlements, property acquired, i.e. structures, houses, wells, number and types of trees on the acquired land. This villagewise information of land use of the land acquired in Raigad district was compiled from the records at Corporate office.

In order to get an idea of the complexity of land use involved within a single category of agriculture at village level, details of agricultural land use of the acquired land in 3 tahsils of Raigad district are given below:

Tahsil	*No. of villages*	*Area acquired in Ha.*	*Type of land use*
Roha	11	66	Agriculture
Mangaon	31	156	Agriculture
Mahad	21	200	Agriculture

Details of agricultural land use

Tahsil						*No.of villages under each land use*							
K	*KR*	*R*	*V*	*VP*	*KRV*	*KPV*	*KPVR*	*KPR*	*RPV*	*KV*	*KP*	*VW*	*NA*
Roha	3	4	1	—	1	—	—	—	—	—	—	—	2
Mangaon	9	—	1	—	11	1	1	2	1	3	1	1	—
Mahad	4	1	—	—	2	—	3	2	—	—	7	1	1

Note: Explanation of land use abbreviations:

K-Kharif, R- Rabi, V-Fallow land, P-Pasture, W-Wasteland, NA- Non-agricultural

Land use Around the Railway Alignment Prior to Construction

The following analysis is based on the interpretation of land use maps on 1 : 250,000 scale compiled by NRSA with the help of satellite imageries of 1988-89.

Coastal Maharashtra

Here the alignment of Konkan railway is in the interior or away from the coast as compared to Goa and Coastal Karnataka. Around Panvel, a variety of land use categories are observed, including agricultural land, both kharif and double cropped to the east, land with or without scrub in the north east, sált affected and marshy land to the north-west and deciduous forest to the south-east and southwards up to Pen. To the north of Roha the region is characterized by deciduous forest with some agricultural land on the bank of Amba river. On the banks of Kundalika river there is agricultural land with Kharif and rabi crops. Beyond cultivated land there is deciduous forest. Moving south towards Kolad and Indapur there is degraded forest and scrub land on both the sides of railway line. Around Mangaon there is agricultural land with kharif crops, but there are some patches of double cropped areas as well. Further south towards Goregaon and Veer scrubland extends on both the sides. Around Karanjadi agriculture (kharif) is the main land use.

In Ratnagiri district agriculture (kharif) and land with or without scrub are the two major land uses around Konkan railway. Around Khed and Chiplun some patched of fallow land are observed. A small patch of semi evergreen forest is observed to the east of National Highway near Aravali road station. Barren, rocky, stony wasteland is found to its south-west. Stony waste interspersed by scrubland is also seen along Bav river around Ukshi and Bhoke stations. The same land use prevails around Ratnagiri and to the east towards Nivsar with isolated patches of agricultural land. There is no change in land use further south from Lanja to Rajapur and beyond. Barren/rocky wasteland and scrub land dominate the landscape. Due to the topographic constraints, a large portion of Konkan railway in Ratnagiri district is under tunnels and the largest tunnel on Konkan railway (Karbude) is also found in this district.

In Sindhudurg district a large patch of semi evergreen forest is located to the south of Vaghotan river and west of National Highway. Scrubland prevails towards the east of railway line. With forest continuing towards the west of Kankavli, some patches of agricultural land with scrub in between are found towards the south and east of Kankavli. A large expanse of scrubland prevails on both the sides of Kudal. Some kharif cultivation and forest land is seen in the Kudal-Sawantwadi section. Good agricultural land with double cropping is found in the valley of Terekhol river to the south and west of Sawantwadi. Further south around Madure some forest and scrub land prevails.

Goa

There is a distinct change in land use around Konkan railway as it enters from Maharashtra to Goa. A large area under plantation is observed around Pernem. Kharif and double cropped area is noticed towards the east. A large patch of deciduous forest is observed to the north of Mandovi river which is bordered by marshy, swampy land on both the sides of Konkan railway. Plantation and swampy land is

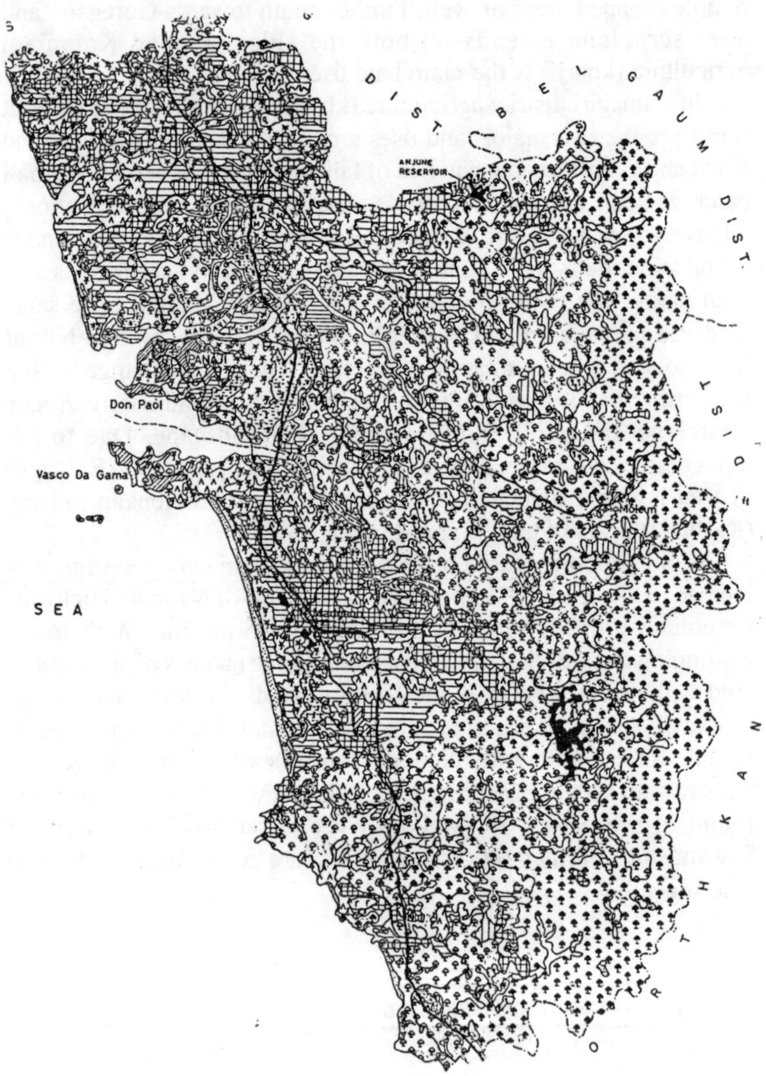

Fig. 3.1 Land Use/Landcover Map of Goa

LANDUSE/LANDCOVER MAP
GOA DISTRICT, GOA

1: 250,000

Km 5 0 5 10 15 Km

REFERENCE

BUILT-UP LAND	Built-up land
AGRICULTURAL LAND	Crop land
	Kharif
	Rabi
	Double crop (Kharif+Rabi)
	Fallow
	Plantation
FOREST	Evergreen/Semi evergreen forest
	Deciduous forest
	Degraded forest/Scrub-land
	Forest blank
	Forest plantation
	Mangrove
WASTELANDS	Salt affected land
	Waterlogged land
	Marshy/Swampy land
	Gullied/Ravinous land
	Land with or without scrub
	Sandy area (coastal & desertic)
	Barren rocky/Stony waste/ Sheet rock area
WATER BODIES	River/Stream
	Lake/Reservoir/Tank, Canal
OTHERS	Shifting cultivation
	Grass land/Grazing land
	Mining areas

BOUNDARY International, State, District

ROAD: Major, others RAILWAY

SETTLEMENT: State Hq., Dist. Hq, Other towns

FOREST Boundary:

also found to the east of Panaji. A mixed land use is found on both the banks of Zuari river-cultivated land with kharif and double cropping, land with or without scrub and plantation. Agricultural land with double cropping is found around Madgaon and kharif crops further south. Deciduous forest and scrubland is observed as Konkan railway approaches Balli and Canacona stations. A patch of evergreen forest is also noticed in between deciduous forest and double cropped area. There are sandy areas along the coast and marshy areas along the river in the extreme south.

Coastal Karnataka

In UK district evergreen forest is the major land use category. However, the alignment of Konkan railway is close to the coast where agriculture and other land use categories are also found. Around Kalinadi creek there is some plantation, kharif crops and evergreen forest. Around Karwar there is some agricultural land with kharif crops, surrounded by evergreen forest and sandy area along the coast. Double cropped area, degraded forest and plantation is observed around Ankola. Kharif cultivation, sandy area and evergreen forest is found to the east of Jadri creek. Plantation and degraded forest is observed around Kumta and Honavar with sandy area along the coast. Similar land use pattern prevails further south up to Bhatkal. More land is under evergreen and degraded forest with small patches of fallow and agricultural land both kharif and double cropped. In DK district plantation and agricultural land, both kharif and double cropped are the major land use categories along Konkan railway. These lands are bordered by scrubland and/or evergreen forest. This land use pattern prevails throughout the coastal belt from Coondapur to Udupi and Udupi to Mangalore. There are sandy areas along the coast. A large area around Mangalore is under plantation and to the south of Mangalore there is an agricultural land with double cropping.

Land use around Konkan Railway after the Construction: A Micro Level Study

This study is conducted with the help of maps on 1 : 50,000 scale for the following selected area around 10 km on both the sides of Konkan railway: (1) Roha in Raigad district of Maharashtra and (2) Panaji and Madgao in Goa.

Land use around Panaji and Madgaon (Goa)

Land use maps for entire Goa state are compiled on 1 : 50,000 scale by the Department of Science, Technology and Environment, Government of Goa. under Integrated Mission for Sustainable Development for the state of Goa and are under publication. Around Panaji and Madgao the Konkan railway is cutting through the Khazan lands interspersed by patches of water logged areas. Dense deciduous forests are observed to the east of railway line. The land use changes to settlement, orchards and crops and also plantations along the National Highway. To the south of Zuari river a variety of land use is observed along the KR alignment as one goes towards Madgao. The railway line cuts across *khazan* lands, forest plantations, agricultural lands and plantations, barren rocky wasteland and dense deciduous forest, a large track of agricultural land, (kharif), settlement, orchards and crops. KR runs parallel to South Central Railway towards east of Madgao. Here settlement, orchard and croplands are followed by agricultural land and plantations.

Land use around Roha (Maharashtra)

A land use map for this region was compiled with the help of satellite imageries available at Maharashtra Remote Sensing Application Centre, Nagpur. The satellite imageries were obtained from IRS 1 C on 25th February 1997. They were superimposed on the topographical maps on 1 : 50,000 scale and land use categories were identified and demarcated by visual interpretation. After the completion of Konkan railway, rapid population growth in Raigad district resulting from urbanization and industrialization is accompanied by changes in land use. The impact of Konkan railway on land use pattern is evident in this micro level analysts. Forest, agricultural land and waste land are the main land use categories in the region. Roha Veer section of Konkan railway was opened for traffic on 27-9-1993. An industrial estate was established near Roha on the south bank of Kundalika river by Maharashtra Industrial Development Corporation (MIDC) prior to construction of Konkan railway. Old and new industrial areas are located to the south of Kundalika river and NH to the east of Roha. Roha tahsil has the maximum forest area (35%) in Raigad dist. The RF between Roha and Kolad consist mainly of teak and cashew trees. Forest cover is noticed between Nagothane and Roha with patches of agriculture (kharif) and scrubland. Within the stretch of NH and Konkan railway in between Roha to Kolad agriculture (kharif) is the main land use. Some patches of scrub land are also noticed. The same land use continues to the north of railway line as well. Mudflats with vegetation are observed to the south of Kundalika river towards the east of Roha. Mangrove swamps are observed at the mouth of Kundalika river. This region has been converted into agricultural and scrub land. To the north of Kundalika river large area is under forest cover. On the toposheet it is marked as Reserved Forest—dense jungle mainly teak. RF dense mixed jungle towards west of railway line is converted into scrub. Grass 2 metre high and dense scrub south of RF to the west of Kolad is now converted into wasteland and scrub. North of railway line open mixed jungle and open scrub is observed. Further south of Kolad the railway line passes through agricultural land, mainly kharif and some rabi. Forest plantation is also noticed within the forest land to the east of Talayli village and to the north of Kolad. Quarrying activity is carried out to the north of Kolad surrounded by forest. On the toposheet this area is shown as stony waste. Wasteland without vegetation is found towards west of Indapur. Wasteland with vegetation (scrub) is also found in between agricultural land in several parts of the region.

Sacred Groves in Konkan

This is a distict feature of land use in Konkan. Sacred groves are developed around the temple of a particular deity in several villages of Konkan. Each sacred grove occupies an area ranging from about 10 to over 100 hectares. These groves are known for the plantation and preservation of a variety of medicinal and rare plant species. No grazing is allowed in sacred groves. Protection of sacred groves is the prime responsibility of sacred grove committees. Encroachments and cutting of trees in sacred groves is considered as an offence and is reported to the tahsildar within three days of the offence committed. The sacred groves were verified physically for their existence in the field by an expert committee of Bombay Natural History Society, Mumbai. They have prepared a village-wise list of sacred groves in coastal districts along with the name of deity and suggested that the protection of sacred groves could only be achieved by notifying them as environmentally sensitive areas under the provisions of Environment Protection Act, 1986. The area under sacred groves in each tahsil as per this list is as follows:

Table 3.5: Sacred Groves in Konkan

District	*Tahsil*	*No. of villages having sacred groves*	*Area in Ha*
Ratnagiri	Chiplun	6	68.28
	Dapoli	4	58.14
	Sangameshwar	9	202.84
	Lanja	1	27.40
	Mandangad	1	18.01
Sindhudurg	Kankavli	6	24.83
	Kudal	5	65.72
	Sawantwadi	3	60.65
	Vengurla	2	20.00
	Vaibhavwadi	4	90.18
Raigad	Alibag	37	842.28
	Mangaon	4	37.01

Source: Report of Bombay Natural History Society on Sacred Groves in Maharashtra, Mumbai, 2001.

Forest Cover

A forest is defined as a plant community predominantly of trees and other woody vegetation. Forest is determined not only by presence of trees but also by absence of other land uses. In India legal connotation of forest is employed. An area of land recorded as forest in revenue records or proclaimed to be forest under a forest law or Act is described as forest. Thus “forest area” is an area recorded as forest in the government records. Forest cover and forest area are two different entities. A land may be recorded as forest area with its management under forest department, but may not have any discernible forest cover. On the other hand, all wooded lands or plantations delineated as forest cover from satellite data may not be legally recorded as forest area as these could be private plantations or institutional wood lots. Some of the forest in our region is under the charge of Revenue Department, Social Forestry and under private ownership. Although majority of forested lands happen to be within legally recorded forest areas, all the changes taking place in the forest cover is not necessarily due to only changes in the forests managed by the forest department. Substantial tree wealth exists in the country in the form of linear plantations along roads, railway lines, canals, etc. Therefore, it is important from policy and planning point of view to know the extent and quality of forest cover within recorded forest areas and outside it. The type of vegetation in a forest is the product of climatic, edaphic and topographic conditions as well as human interface. The forest cover is classified into two classes by FSI as dense (with crown/canopy density over 40%) and open—all lands with a forest cover of trees with a crown/canopy density between 10 to 40 per cent. The recorded forest area is categorized into Reserved, protected and unclassed forest. Forests are classed as evergreen, deciduous and degraded in NRSA's statistics of land use and land cover.

The State of Forest Report (SFR) 2001 provides much more comprehensive status of forest and tree cover in the country than the previous seven reports. All lands with discernible tree canopy, irrespective of species, land use and ownership have been included in forest cover assessment. For the 2001 assessment of forest cover digital technique instead of visual method was used at a scale of 1 : 50,000. Significant extent of area outside continuous wooded lands, excluded from forest cover assessment earlier has been included in 2001 assessment. Certain areas with tree canopy such as coconut, orchards, coffee and tea estates, etc. which were not taken into account during forest cover assessment earlier, have been included in the present assessment. The current assessment therefore provides a truer picture of the status of forest cover. Survey of India toposheets show forest areas with double dotted lines and wooded lands as green wash patches. These are usually based on surveys done several years back and may be at variance with current status of forest boundaries.

Extent of forest cover is a matter of concern throughout the country. Indian forests are under immense pressure to supply forest products to a growing population in a land scarce situation. Deforestation and degradation is a subject of growing concern in India with far reaching environmental, social and economic consequences. The experience of development projects in our country reveals that infrastructure development is generally associated with deforestation or decline in area under forest. When statistics of area under forest as given by different sources for two different dates, i.e. prior to construction of Konkan railway (1989-90) and after the construction of Konkan railway are compared for our study region at tahsil, district and state levels, a clear picture of increase in forest area emerges, pointing out that construction of Konkan railway has no adverse impact on the forest cover of this region.

Forest Cover in Tahsils through Which Konkan Railway Passes

The Konkan railway passes through 28 tahsils of 7 districts of 3 states:

Maharashtra, Goa and Karnataka. There is no forest cover in 3 out of these 28 tahsils in 1998-99-Ratnagiri, Salcete and Mormugao.

Table 3.6: Changes in Area Under Forest in Tahsils through which Konkan Railway is Constructed

Tahsil	*Area under forest in ha*		*Area under forest*			
	1989-90	*p.c. to tahsil area*	*1998-99*	*p.c.*	*Increase*	*Decrease*
Roha	21,881	34.62	22,600	35.76	719	
Mangaon	13,255	14.11	13,300	13.48	45	
Mahad	8,371	10.31	9,100	11.23	729	
Khed	2,322	2.27	3,700	2.98	1,378	
Chiplun	377	0.31	200	0.26		177
Sangameshwar	142	0.11	0	0		142
Ratnagiri	0	0	0	0		
Lanja	1,033	1.4	1,000	1.3		33
Rajapur	54	0.04	800	0.65	746	
Vaibhavwadi	2,114	5.08	2000	4.8		114
Kankavli	14,628	19	9200	11.9		5,428
Kudal	12,823	15.75	9800	11.96		3,023
Sawantwadi	12,677	9.38	12,900	9.61	223	
Pernem	1,387	5.51	1319	5.45		68
Bardez	4	0.01	0	0		4
Tiswadi	—	—	178	1.07	178	
Mormugao	—	—	0	0		
Salcete	—	—	0	0		
Quepem	7,641	27.14	11,679	33.62	4,038	
Canacona	12,218	36.65	14,328	41.25	2,110	
Karwar	54,337	74.22	55,427	75.7	1,090	
Ankola	72,900	79.35	75,474	82.04	2,574	
Kumta	39,194	67.19	39,687	68.03	493	
Honnavar	57,420	76.07	57,632	76.35	212	
Bhatkal	25,295	72.49	25,433	72.89	138	
Coondapura	67,144	46.5	62,605	40.3		4,539
Udupi	6,322	7.6	4,686	5.58		1,639
Mangalore	0	0	2,703	3.17	2,703	
Total	*433,542*		*435,751*		*17,376*	*15,167*

Ankola in Uttar Kannada district records the maximum (82%) forest area. In Raigad district forest cover is significant only in Roha tahsil (35%) compared to other two southern tahsils. Due to elongated shape of Ratnagiri district, Konkan railway passes through its 6 tahsils where the forest cover is extremely limited from nil to (3%). Half of the district's forest is concentrated in Khed tahsil. In Sindhudurg district forest cover ranges from 5 to 12 per cent in its four tahsils. In

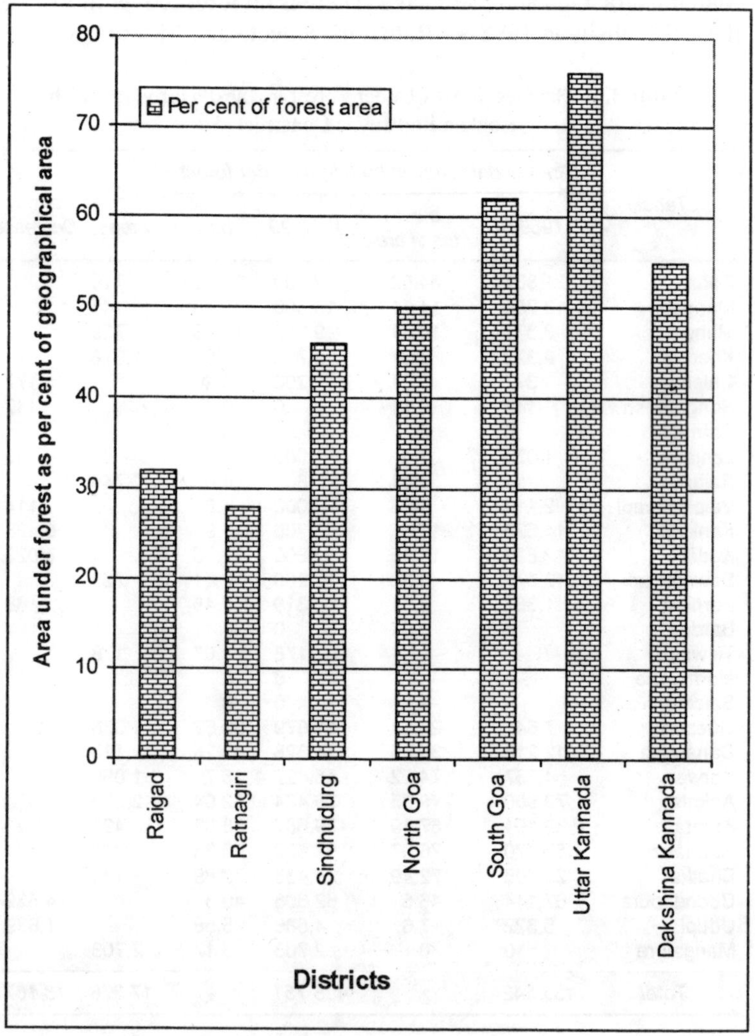

Fig. 3.2 Area under Forest, 2001

Goa 4 tahsils have negligible forest. Pernem in North Goa has a limited forest cover (7.31%) while Quepem and Canacona tahsils of South Goa have considerable forest (over 34 and 41% respectively). In Karnataka Konkan railway passes through 5 coastal tahsils of Uttar Kannada. Area under forest ranges from 70 per cent in Kumta to

82 per cent in Ankola tahsil. In Dakshina Kannada district forest cover is substantial only in Coondapura tahsil (40%). It is limited in Udupi (6%) and Mangalore (3%) in the extreme south.

Changes in Forest Cover at Tahsil Level

Table 3.6 presents area under forest in tahsils through which Konkan railway is passing as per cent of tahsil area and increase/decrease in forest cover since the construction of Konkan railway (in 1989-90). The total forest cover in tahsils through which Konkan railway passes comprised of 433,542 ha prior to construction of Konkan railway (in 1989-90), which increased to 435,751 ha. in 1998-99. Increase in forest cover by 17,356 ha. is recorded in 15 tahsils, whereas in 10 tahsils the extent of forest cover declined by 15,167 ha. However, an actual increase of 2209 ha. of forest area is recorded Three tahsil—Ratnagiri, Salcete and Mormugao which were devoid of forest prior to construction of Konkan railway have no forest cover even thereafter. The maximum increase in area under forest at tahsil level (4038 ha) is recorded at Quepem of south Goa district while the largest decrease of 5428 ha is recorded at Kankavli tahsil of Ratnagiri district. Due to this the per cent of forest in Kankavli declined from 19 to 12 per cent Among the 13 tahsils of Maharashtra area under forest increased in 6 tahsils and declined in 6, recording a net decline by 5077 ha. In Goa with the increase in forest cover in 3 tahsils and decline in 2, the total forest cover increased by 6254 ha. In Karnataka the forest cover increased in all the 5 tahsils of Uttar Kannada district. In Ankola tahsil the percentage of forest in the region increased from 79 to 82 p.c., the highest in the region. In Dakshina Kannada district Mangalore tahsil which had no forest cover in 1989-90 records 3 per cent of its area under forest in 1998-99 with a net gain of 2703 ha. The other two tahsils recorded a significant decline.

Changes in Forest Cover at District Level

Forest cover in our study region at the beginning of Konkan railway construction and in the year 2001 is given below in Tables 7 and 8 and illustrated in Figs 3.1.

Table 3.7: Type of Forests along the Western Coast, 1988-89

(Figs in sq. km.)

District	Evergreen forest	Deciduous forest	Degraded forest	Forest plantation	Mangrove forest*	Total forest area
Raigad	423	525	483	0	49	1601
Ratnagiri	92	11	0	0	2	105
Sindhudurg	487	629	39	5	0	1159
Goa (N&S)	794	1201	20	65	4	2109
Uttar Kannada	6442	1011	625	125	0	8202
Dakshina Kannada	2095	95	12	12	0	2215

Notes: *An additional category of Forest Blanks is noted in Raigad (121 km) and Goa (25 km).

Source: Report on area statistics of land use/land cover generated using Remote Sensing Techniques, NRSA-1995.

Table 3.8: Forest Cover along the Western Coast, 2001

(Figs. in sq. km.)

District	Geographical area	Dense forest	Open forest	Total	%	Scrub	Mangrove
Raigad	7152	1214	1074	2288	31.99	375	34
Ratnagiri	8208	1905	306	2211	26.94	279	9
Sindhudurg	5207	1970	408	2378	45.67	109	1
North Goa	1736	748	121	869	50.06	2	4
South Goa	1966	1037	189	1226	62.36	14	1
Uttar Kannada	10291	6502	1305	7807	75.86	1	—
Dakshina Kannada	8441	3524	1193	4717	55.88	2	2

Source: State of Forest Report 2001, published by Forest Survey of India, (Ministry of Environment and Forest) Dehra Dun.

Table 3.9: Changes in area Under Forest

District	As per Director of Agriculture			As per Forest Survey of India		
	1989-90	1997-98	Change	1989-91	1999-00	Change
Raigad	1526	1680	+154	1650	2288	+638
Ratnagiri	45	67	+22		2211	
Sindhudurg	586	530	-56		2378	
Rat & Sind.				5621	4589	-1032
Goa	896	1255	+359	1247	2095	+848
Uttar Kannada	7824	8152	+328	7813	7807	-5
Dakshina Kannada	2330	2275	-55	3640	4717	+1077

Source: District Census Handbooks of respective districts, 1991
District Statistical Abstracts of respective districts, 1998-99
Forest Survey of India, State of Forest Reports, 1993 and 2001.
+ Increase, –Decrease.

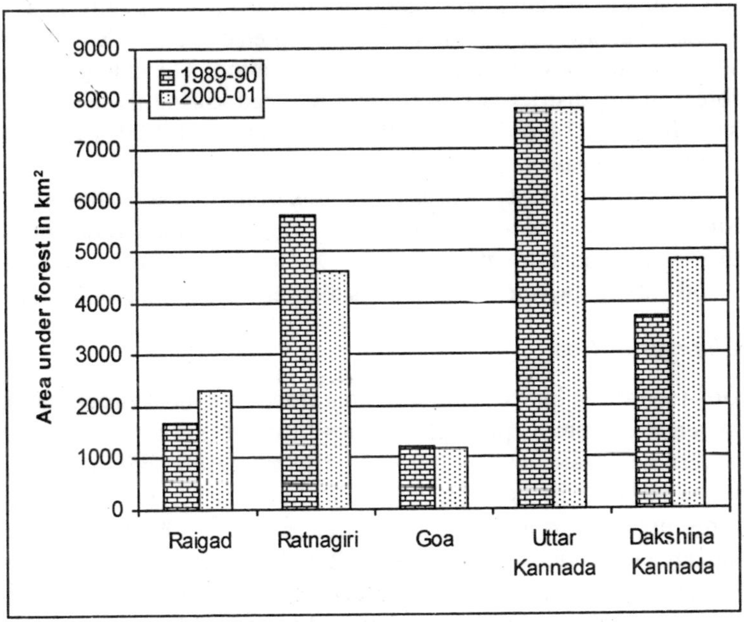

Fig. 3.3 Changes in Area under Forest

Changes in Coastal Maharashtra

A comparative study of FSI statistics at district level reveals an increase in forest cover in Raigad district. Area under both dense and open forest has increased (Toposheets 47 F&B, E.G.). The State of Forest Report 1999 notes that more area under mangrove changed to non-forest, while from non-forest some area changed to mangrove. There is an overall decline in area under mangrove forest. The actual increase in forest area as per FSI is more (638 km^2) compared to an increase of 154 km^2 recorded by Directorate of Agriculture. Up to 1999 FSI published the statistics for Ratnagiri and Sindhudurg districts together (undivided Ratnagiri district). The combined statistics of Ratnagiri and Sindhudurg districts for 2001 reveal a decline in forest area by 1032 km^2, and in percentage to district area from 43 to 34.2. There is an increase in dense forest and mangrove forest and decrease in open forest (Toposheets 47L,48 I). According to Director of Agriculture, area under forest in Ratnagiri district has increased by 22 km^2, while loss of forest cover is noticed in Sindhudurg district (by 56 km^2). The vast difference in the two data sets can be attributed to visual and digital interpretation of satellite data. The hilly areas of Ratnagiri district appear to have been interpreted as under dense forest, while in fact they are devoid of any vegetation. The extent of open forest in the district is significantly reduced during the short time span from 3321 km^2 in SFR 1999 to 714 km^2 in SFR 2001 probably due to same reason.

The 1981 District Census Handbook of undivided Ratnagiri and Sindhudurg dist notes that there are extensive Malki forest in the district, i.e. under private ownership. Due to unlicensed and unlimited wood cutting, private forests in this district were gradually denuded.

Changes in Goa

Goa records an increase in forest area from both the sources, but the increase is more as per FSI statistics (848 km^2) compared to State government statistics (by 359 km^2). The increase in FSI statistics is witnessed only in the latest 2001 SFR. In all earlier assessments the forest area in Goa has remained unchanged. SFR 2001 records 1785 km^2 area under dense forest, while 1999 SFR record 995 km^2 as dense forest. Area under open forest increased from 251 to 310 km^2. The percentage of area under forest in Goa increased from 22.8 to 43.5 per cent The changes have taken place in Toposheets 48 E/15 and E/15. The statistics of North Goa and South Goa are separately published in SFR 1999 and 2001. South Goa had a much larger area under forest (43.49%) compared to North Goa district (22.81%) in 1999. The percentage has increased in both the districts in 2001. The increase in area under dense forest in North Goa district (over about threefold) is more prominent compared to South Goa district. A remote sensing based study of selected mining region of Goa undertaken by TERI shows that vegetation cover has reduced by more than half due to intensive mining of the area over a period of 10 years. But large scale afforestation work has also been carried out under various schemes covering 12,732 ha of area during 1990-91 to 1998-99.

Changes in Coastal Karnataka

In Uttar Kannada district there is an increase in forest area (by 328 km^2) as per State government statistics, while decrease in forest area (by 1 km^2) as per FSI sources. Here dense forest declined by 873 km^2 and appears to have been converted into open forest, as open forest increased by 882 km in SFR 2001 compared to SFR 1999. In earlier assessments, no such change is noted. In the field studies of this district loss of trees was observed in all the study sites. After 10 years there was a decline in total number of trees in forest sites of evergreen forest zone. In moist deciduous forest zone there was an increase in the number of species. As per NRSA data 125 km^2 of area was under forest plantation in 1988-89, which can explain part of the increase in open forest. Dakshina Kannada district records a loss of 55 kms of forest cover as per Director of Agriculture. On the contrary, the gain in forest cover by 1068 km^2 is seen as per FSI. Area under dense forest decreased by 759 km^2, open forest increased by 311 km, and mangrove forest remains unchanged. This increase is confined to last two years (SFR 1999 to 2001). Decline in forest area in tribal districts causes more hardships to vulnerable section of society. On this background, registering an increase in forest area of a tribal district of South Kannada is significant.

Summary of Changes

According to Directorate of Agriculture of Maharashtra, Goa and Karnataka states, area under forest in our study region in 1989-90 was 13,206.61 km^2. In 1997-98 it was found to be 13,958.53 km^2, recording an increase of 751.92 km^2. Forests constituted 31.24 per cent of our region's geographical area prior to construction of Konkan railway, and this increased to 33.02 per cent after the completion of its construction. According to FSI area under forest recorded an increase of 1526 km^2 from 19,971 km^2 in 1989-91 to 21,497 km^2 in 1999-2000. The per cent of area under forest increased from 46.84 per cent to 50.42 per cent during this period. More remarkable is the increase in area under dense forest from 13,681 km^2 to 16,900 km^2. Dense forests constitute 78.61 per cent of total forests in 1999-2000 compared to 68.5 per cent in 1989-91.

The statewise changes in forest cover (confined to study region) are as follows:

	Forest Survey of India	*Directorate of Agriculture (figs. in km^2)*
Maharashtra	-394	+120
Goa	+848	+359
Karnataka	+ 1052	+273

The above statistics points out that infrastructure development, i.e. construction of Konkan railway is not associated with deforestation. In fact, the statistics reveal a gain in forest cover on the Western coast of India. This gain is of 1526 km^2 as per FSI and 752 km^2 as per Directorate of Agriculture. Konkan railway has not adversely affected the forest cover in this region, What is remarkable as per SFR is the increase in area under dense forest by 3219 km^2 while decrease in open forest by 1587 km^2. The gain in forest cover in our study region is the outcome of stringent rules and regulations for the acquisition of forest land being implemented along with various schemes of new plantation and afforestation projects undertaken such as social forestry and wasteland development. The social Forestry Departments of respective state governments made consistent efforts for massive tree plantation under various schemes. Tree plantation was undertaken by KRCL in the premises of railway stations, staff colonies, along the tracks. The details of afforestation are discussed in Chapter 8.

Spatial Distribution of Other Categories of Land Use

Agricultural Use

Our country being predominantly agricultural, cultivation is practiced since long ago. With population growth and increase in population density farming has been extended on marginal lands and on slopes of hills as no good quality land is left for further expansion. This is true for every village on the western coastal region as well. The major forms of environmental deterioration associated with population growth include deforestation, soil degradation, watershed destruction, and urban encroachment on prime agricultural land. Land under agricultural use constitutes 34 per cent of our region's geographical area. Net area sown (NAS) is the largest component of land under agricultural use and comprises of 60 per cent of total agricultural land. The extent of NAS depends on the physiography of the region. Due to the limitations imposed by the Western Ghats in the east and the Arabian sea in the west, NAS occupies one fifth (20.33%) of the region's geographical area. An overwhelmingly large (86%) proportion of land under cultivation is under kharif crops and the remaining under rabi crops. Double cropped area constitutes only 7.8 per cent of net area sown and is smaller in extent than the area under rabi crops. There are tremendous variations at district level in the proportion of net area sown ranging from the maximum of 37.9 per cent in Ratnagiri district to the lowest of 9.1 per cent in Sindhudurg district. Double cropped area constitutes the highest of 36.31 per cent of NAS in Goa to the lowest of less than one per cent (0.8) in Ratnagiri district. Fallow land comprises one-fifth (20%) of total agricultural area. The proportion of fallow land is the highest (22% of district's area) in Ratnagiri, while no fallow land is recorded in Goa. The absence of fallow land in Goa can be attributed to more land devoted to tourism related activities. Land used for tourism related activities is not separately classified in land use statistics and is not available from any other source either. Changes on account of tourism are therefore difficult to analyze.

The western coastal region is known for its agricultural plantations and these plantations constitute over one fifth (21%) of the region's agricultural area. This proportion reaches the maximum of 53.6 per cent in Dakshina Kannada district, closely followed by Sindhudurg district where half (50.3%) of agricultural area is under plantation. DK district has the plantation of cocoanut, arecanut, rubber, pepper, cashew, whereas Sindhudurg district is known for the plantation of Alphanso mangoes, which are exported in large quantities. Goa has one-third of agricultural land under plantation mainly cashew nut and cocoanut. In Raigad dist area under plantation is insignificant (1.1%). Under cropland, kharif crops dominate, whereas double cropped area (kharif + rabi) constitutes 13 per cent of agricultural land. In Goa only 3 per cent of agricultural land is fallow.

Wasteland

Wastelands are those land areas which are not used to the level of their optimum potential but can be brought under vegetation cover with some efforts. Wasteland is defined as "that land which is degraded and is presently lying unutilized except as current fallow due to different constraints." Wastelands can result from inherent and/or imposed disabilities such as erosion, degradation, salinity, water logging, etc. The wasteland information when used in conjunction with data on other resources help in optimum utilization of wastelands for agriculture, afforestation, etc. thereby reducing the pressure on arable land. Information about wasteland enable planners and decision makers to prepare various programmes for reclamation, management and development of such lands more productively. The large extent of wasteland directly contributes to poor productivity, low per capita income, increasing pressure of population on land, continuously reducing per capita availability of food, unemployment and malnutrition especially in rural areas. The major ill effects of wasteland are ecological imbalances, floods, droughts, poor recharging of underground water, drinking water scarcity. There is a need for taking urgent and progressive steps for putting wastelands to productive use.

In order to have a reliable data base on the type, extent and location of wastelands, mapping of wastelands in the country was completed under National Wasteland Inventory Project using remote sensing technology. This enabled to priorities about 180 districts in the country which are critically affected by the problem of wasteland. Such districts which have large areas of wasteland (over 15% of geographical area of the district) were identified.

According to the statistics published by NRSA in the Wastelands Atlas of India and as per the Report on area statistics of land use and land cover generated using remote sensing techniques by NRSA, 18.9 per cent of the region's geographical area is classed as wasteland. The statistics point out the magnitude of the problem of wasteland reclamation in the region. The wasteland statistics by these two sources are given in the table. According to Wasteland Atlas,10 categories of wasteland are found in our study region, whereas 5 categories of wasteland are noted in Report on area statistics of land use/land cover. The area occupied under each category in our study region is as follows:

Table 3.10: Area under Wasteland by Categories

	Category	*Area in sq. km as per Wasteland Atlas*	*Land use/land cover st.*
1.	Land with or without scrub	4916.94	5995.57
2.	Waterlogged/marshy, swampy land	284.88	284.40
3.	Land affected by salinity/alkalinity	173.16	269.64
4.	Underutilized/degraded notified forest land	826.78	—
5.	Degraded pastures/grazing land	2.47	—
6.	Degraded land under plantation crop	687.11	—
7.	Sands—inland/coastal	54.38	118.55
8.	Mining/Industrial wastelands	112.89	—
9.	Barren, rocky/stony waste/sheet rock area	704.00	1370.27
10.	Steep sloping area	69.57	—
	Total wasteland	7832.18	8038.48

Source: Report on area statistics of land use/land cover, NRSA, 1995.
Wasteland Atlas of India, NRSA, 2000.

The statistics of wasteland at district level are given below in table:

Table 3.11: Wasteland as per cent of District's Area

District	*As per Wasteland Atlas*	*As per Report on Land use*
Raigad	30.18	24.67
Ratnagiri and Sindhudurg	26.45	—
Ratnagiri	—	29.52
Sindhudurg	—	51.34
Goa	—	9.48
North Goa	18.50	—
South Goa	14.87	—
Uttar Kannada	7.41	3.92
Dakshina Kannada	10.01	5.35
Average, all districts	18.37	18.85

Source: Report on area statistics of land use/land cover, NRSA, 1995.
Wasteland Atlas of India, NRSA, 2000.

Wasteland constitutes the largest proportion of geographical area (51.34%) in Sindhudurg district. Uttar Kannada has the lowest proportion of wasteland (3.92%) among all districts.

Among 10 types of wastelands the most widespread type found in our study region is the land with or without scrub. Due to thin soil cover in the Western ghats, over two third of the total wasteland is classed as scrubland. Extent of scrubland according to Land use statistics is much more than the wasteland statistics in Ratnagiri, UK and DK districts. This is the most prevalent category in Ratnagiri and Sindhudurg districts constituting over two-third (68.5%) of area under wasteland in these districts. The extent of scrubland in these districts occupies about half (48.13%) of the region's total scrubland. This is closely followed by Raigad district, where the scrubland occupies another one third of the region's total. The extent of scrubland is limited in Goa and coastal Karnataka.

Next in importance to scrub is underutilized/degraded forest. This category constitutes one-tenth of total wasteland. Those lands with various types of forest cover in which vegetation cover is less than 20 per cent are classified as "degraded forest". As per the Wasteland Atlas, degraded forest constitutes 1.95 per cent of the region's geographical area. This is the most prevalent type in Uttar Kannada district where it constitutes 77 per cent of district's total wasteland. The degraded forest in UK district constitutes 62 per cent of the region's total degraded forest. Raigad and UK districts together comprise 94 per cent of the region's degraded forests.

Degraded land under plantation crops constitutes 8.77 per cent of total wasteland in the region. Over 90 per cent of it is concentrated in Ratnagiri and Sindhudurg districts as land under plantation crops is also found in these districts. This category constitutes 18.16 per cent of total wasteland in these districts, whereas in Goa, it comprises 5.25 per cent of wasteland. The extent is more in North Goa compared to South Goa district. There is no land under this category in UK district and it has a limited extent in DK district. Degraded pasture/grazing land has a limited extent in South Goa district. The excessive grazing has converted most of the community grazing lands into wasteland. Under Wasteland Development Programme it is proposed to develop degraded pasture land by adopting improved techniques to meet increasing demand for fodder for cattle.

Barren, rocky/stony waste/sheet rock area is widespread over the Western Ghats. Soil cover is more or less absent. Hence these areas are devoid of any vegetation. This category is widely prevalent in Raigad, Ratnagiri and DK districts. As per Wasteland atlas, this category constitutes 1.61 per cent of geographical area, whereas according to land use statistics, it comprises 3.24 per cent The two sets of data differ significantly at district level. In Goa, UK and DK districts the extent as per Wasteland Atlas is far more than the other source, whereas in Ratnagiri and Sindhudurg districts the reverse is the case. As per Wasteland Atlas, this category constitutes 2.21 per cent of the district's geographical area, and as per Land use data, it comprises of 9.08 per cent According to land use statistics, 86 per cent of the region's barren land is concentrated in Ratnagiri and Sindhudurg districts, whereas according to Wasteland Atlas, the extent of this category in Ratnagiri and DK districts is more or less the same and comprise about 40 per cent of the region's wasteland.

The extent of waterlogged and marshy land is less than one per cent As per wasteland statistics, it constitutes 0.88 per cent of geographical area, whereas it constitutes a little less (0.67%) as per the other source. Marshy/swampy land is found in all the districts except Sindhudurg, with the largest extent in Raigad district. In all the districts except Goa, the extent of marshy land is much more as per land use statistics, compared to Wasteland Atlas statistics.

About half of the region's waterlogged land is found in Raigad district alone. In Goa, marshy land constitutes 3.44 per cent Land affected by salinity/alkalinity is found localized only in Raigad district. It constitutes 2.42 per cent of the district's geographical area. Salt affected area in Raigad district comprises of 3.46 per cent of district's area. As per the Wasteland Atlas, no other district of the region has this category of wasteland, whereas land use statistics point out that this category is also found in Ratnagiri district (0.27% of district area). Mining/industrial wasteland is confined to Goa. Wasteland under mining is concentrated mainly in North Goa. Small patches of mining wasteland are also found in Sindhudurg and UK districts. Steep sloping area is confined to Ratnagiri and Sindhudurg districts and to some extent in Goa. Sands-inland/coastal have a limited extent in all districts except Goa.

Built up Land constitutes an average of 1.97 per cent of the region's geographical area and is recorded in all the districts. The statistics of built up land published by NRSA and by MRSAC differ considerably as shown below:

Table 3.12: Built up Land

(*Figs. in ha*)

Source	*Raigad*	*Ratnagiri*	*Sindhudurg*
NRSA	17235	28334	6993
MRSAC	1126	8040	1656
Land under water bodies			
NRSA	50803	14795	8029
MRSAC	50803	593	8500

Source: (1) Report on area statistics of land use/land cover, NRSA, 1995.
(2) Natural Resource Atlas of Raigad, Ratnagiri and Sindhudurg districts, (Maharashtra State), MRSAC, Nagpur, 1994.

As per NRSA, the largest extent of built up land is found in Ratnagiri district. This district onstitutes over one third (34%) of the region's built up land. Raigad district ranks next with one fifth (20.68%) of the region's built up land followed by DK district (18.5%). The lowest proportion of built up land is found in Uttar Kannada district. The difference between the two sets of data in Raigad and Sindhudurg dists is difficult to account for. Comparable data for other dists is not available.

Land Under Water Bodies comprises all the land under rivers, streams, lakes, reservoirs, canals. Together, this category comprises

2 per cent of the region's area. Among the water bodies, land under rivers and streams occupy over 90 per cent of district's area in all districts except Uttar Kannada. In UK district land under lakes, reservoirs and canals is much more compared to other districts, and comprise half of the district's area under water bodies. Water bodies in the form of rivers, lakes, reservoirs, tanks occupy 5 per cent of Goa's total geographical area. Goan coast is full of creeks and estuaries which provide a shelter to fishing crafts. Like built up land, difference in the statistics of water bodies for Ratnagiri district is remarkable. No such difference is notice in other districts.

Salt Pans: Land under salt pans has a limited coverage and is confined to two districts—Raigad and Uttar Kannada As per DST Goa's statistics area under salt pans is restricted to 11 km^2 in Goa (0.3%).

Khazan Lands: A special feature of Goan topography is the *Khazan* land. The *khazans* are saline flood plains in Goa's tidal estuaries (below sea level at high tide) which have been reclaimed over centuries with an intricate system of bunds (dykes and sluice gates). This land is confined to 2.27 per cent of the state's area, and is under various pressures all over Goa.

Mining Area comprises 0.1 per cent of the region's area and is confined to Goa, Sindhudurg, UK and DK districts. Due to availability of iron ore mining area in Goa constitutes 0.88 per cent of geographical area. As per DST Goa's statistics land under mine pits and mining areas constitutes 2.6 per cent of total area. Other districts have a very small extent of mining area due to lack of mineral resources.

Grass/Grazing Land is confined to Raigad and Goa. In Raigad dist. it constitutes 0.5 per cent of geographical area. The small extent of grazing land is due to utilization of marginal lands and slopes of hills which were used as common grazing lands for cultivation. The expansion of cultivation exposed slopy lands to erosion and reduced the availability of grazing land.

Settlement with Mixed Vegetation: This land use category is found only in one district, i.e. DK. where it comprises of 28.6 per cent of district's geographical area. This category comprises of 5.7 per cent of the region's area. A special features of Goan landscape is settlement + orchard and another is settlement + orchard + crop. These two categories together occupy 9.66 per cent of Goa's geographical area. In fact, if this category is added to agricultural land, it increases substantially.

Changes in Land use after the Commissioning of Konkan Railway

Land use after the completion of Konkan railway is studied at district level with the help of latest statistics available for the years 1998-99 or 1999-2000, published by the Bureau of Economics and Statistics (BES) of respective state governments. These land use categories are not comparable with the remote sensing data prior to construction of Konkan railway. For a comparative study of land use prior to construction of Konkan railway, statistics published by Bureau of Economics and Statistics for the years 1991-92 are used. The per cent of land under each category for both the years are shown below.

Table 3.13: Land use Changes

Land use category	*% of geographical area*	
	1991-92	*1998-99*
Land not available for cultivation*	18.87	16.97
Permanent pastures and other grazing land	2.40	3.00
Land under miscellaneous tree crops and groves	3.25	5.40
Cultivable wasteland	12.33	9.83
Net area sown	23.57	24.51
Fallow land	6.75	8.01

* Includes land put to non agricultural uses such as buildings, roads and area under water, barren and uncultivable land.

Source: Bureau of Economics and Statistics of Government of Maharashtra, Goa and Karnataka.

Net area sown (NAS) comprises about one-fourth of the region's total area. It varies at district level ranging from the highest of 50 per cent in Goa to the lowest of 11 per cent in UK district. There is a vast difference in these statistics from remote sensing. NAS as obtained from remote sensing is much more than from BES in Raigad and Ratnagiri districts. In rest of the districts, NAS as per remote sensing is less. The difference is most remarkable in case of Sindhudurg district (9.1% as per remote sensing, 26.9% as per BES). The table reveals that NAS in the region has increased over the years. The clearing of new land for agricultural use is a response to pressures on the land, increasing land fragmentation, and lack of adequate non-farm rural employment opportunities. The increase in NAS is substantial in Ratnagiri, Sindhudurg and North Goa districts, marginal in South Goa and DK. There is an increase in plantation and agricultural areas. As per TERIs, report (2002) NAS in Goa has increased during this period because of the cultivation of cultivable wasteland (This includes fallow land). In Ratnagiri district net area sown increased from 28.6 per cent to 31.8 per cent of geographical area. In DK districts it has declined. The decline is remarkable in Raigad dist. from 26.9 per cent to 23.2 per cent of district's area due to urbanization and industrialization. Area sown more than once is very small, constituting only 0.12 per cent of NAS in the region. This is an indication of lack of infrastructure facilities like assured irrigation in the region. In terms of hectares, the extent is the largest in DK district and the smallest in Ratnagiri district. Area sown more than once has declined in all the districts except Goa.

***Land under Miscellaneous Trees and Groves*:** is identified as a separate land use category. It follows the same increasing trend like NAS. Land under this category in Raigad and UK district remains unchanged, while in DK district it has declined. The increase in land under groves is confined to two districts, Ratnagiri and Sindhudurg. In these districts land under this category increased from less than 3 per cent to 6.18 and 12.63 per cent of total area respectively. The increase in these districts can be attributed to various schemes implemented under Horticulture Development Programme by the state government. These schemes include plantation of forest species for industrial and commercial uses, development of minor forest produce, massive afforestation programme, afforestation for soil and moisture conservation on special sites, afforestation of degraded forest under tribal sub-plan, Western Ghat Development, Eco-development and eco-conservation programmes, compensatory plantation.

***Land not Available for Cultivation*:** Due to topographic and other constraints in the coastal region, land not available for cultivation constitutes about 17 per cent of geographical area. On district level, it varies from the lowest of 4.69 per cent in UK district to the maximum of 26.11 per cent in Sindhudurg district. It is apparent that with the increase in NAS, land not available for cultivation in our study region has declined. However, the decline is confined to 3 districts only. In Raigad and UK districts this decline is associated with decline in NAS and increase in cultivable wasteland. In Ratnagiri district the decrease is associated with the increase in NAS and cultivable wasteland. In North and South Goa and DK districts both the land not available for cultivation and NAS has increased. In Goa, this category is inclusive of land under mining and tourism activities.

There is no separate data available on land use changes due to mining and tourism activities. Of the 11 tahsils in Goa, 5 are in the mining belt. According to the Report by TERI (2002), there are large scale land use changes taking place in Goa due to mining and quarrying activities. The land cover analysis of mining areas in Goa carried out by TERI with remote sensing techniques reveals that there was large scale expansion of mining activity in study villages over a period of 10 years. Mine pits and dumps have shown highest growth in terms of area. A change in land use from agriculture or forest to quarrying is also observed. Quarrying for laterite stones and aggregates is very common in Goa. Land use statistics of Agriculture Department fail to account these changes due to mining activity.

Tourism is the major economic activity in coastal talukas of Goa. Bardez has 23 per cent of area under land not available for cultivation. Substantial part of this is transformed to constructed area to meet the demand from tourism industry and relatively small area is found under barren category. Tiswadi and Ponda talukas have also registered a higher growth in non agricultural lands. Major land use changes are taking place in coastal villages and vulnerable ecosystems in the villages—sand dunes, mangroves and *khazans* are experiencing extensive damage (TERI,2000 b). The statistics of tourists arrivals in each tahsil over the years and their increasing trend point out the land conversion (land put to residential and non residential use). Increase in number of hotels, guest houses, and other facilities for tourists is also evident from Goa's Accommodation Directory published by the Directorate of Tourism. It provides an indirect evidence of changes in land use in Goa. Konkan railway has helped in boosting tourist traffic in Goa.

Fallow land is a temporary phenomenon, which varies from year to year. It occupies 8 per cent of the region's area. Ratnagiri district records largest extent (14.5% of district area). Goa does not record any fallow land. Fallow land has increased in all the districts except DK where it has declined. Permanent pastures and grazing land constitutes 3 per cent of the region's land, which has marginally increased over the years. Land under pastures increased in Ratnagiri, Sindhudurg and UK and declined in Raigad and DK dists. It has a limited extent in Goa.

Cultivable Wasteland: The area under cultivable wasteland in the region has declined from 12.33 to 9.83 per cent The decline is confined to 4 districts—Sindhudurg, North and South Goa and DK.

In UK district there is a small increase, whereas in Raigad and Ratnagiri districts it has substantially increased. The decline in cultivable wasteland is associated with increase in NAS. However, a substantial increase in cultivable wasteland in Raigad district is matched by decrease in NAS. Ratnagiri district records increase in both the categories. In Goa the decline in cultivable wasteland can be attributed to increase in land used for tourism related activities.

Land use Changes at Tahsil Level

Tahsil level changes in area under forest are already discussed in part II of this chapter. Tahsil wise statistics of other land use categories for these tahsils prior to construction of Konkan railway are now compared with the land use after the construction of Konkan railway. There are three other land use categories comparable at this tahsil level. Net area sown (NAS) in 1997-98 constitutes 29.7 per cent of total tahsil area, whereas cultivated land, irrigated and unirrigated constituted 31.5 per cent in 1989-90. This shows that aggregate area under cultivation in these tahsils has declined, though there is an overall increase in net sown area in our study region. NAS has declined in 13 tahsils, increased in 11 tahsils and remained unchanged in 4 tahsils. The increase in NAS is found in all the tahsils of Goa and DK, whereas decline in NAS is observed in all the tahsils of Raigad, 3 tahsils each of Ratnagiri and Sindhudurg and all the tahsils of UK dist. Double cropped area constitutes 12 p. c. of NAS in 1997-98. Land not available for cultivation constituted 22.88 per cent of tahsils' area prior to construction, which declined to 18 per cent upon completion. The area declined in 17 tahsils, increased in 6 and remained unchanged in 5 tahsils. In general, it has been observed that the tahsils which experienced decline in NAS also recorded decline in area not available for cultivation. However, this is not the case in Goa. Here, increase in NAS is associated with decline in area not available for cultivation. Uncultivable land or culturable wasteland declined from 22.3 per cent to 21.46 per cent Like the other two categories, the tahsilwise analysis reveals variations, decline in area is recorded in 10 tahsils, increase in 12 and no change in 6. But it cannot be associated with the other two categories. Statistics of fallow land are not available on tahsil level prior to construction. The latest statistics reveal that 11.12 per cent of total land is fallow. The extent of fallow land is more in tahsils of Maharashtra state, and much less in coastal Karnataka, while there is absence of fallow land in Goa.

Conclusion

Barren/rocky land is the predominant land use in Konkan. Maximum land acquired for construction (42.4% of total) was barren land, followed by agricultural and garden land (35.4%). Forest land constituted only 6 per cent of the total land acquired. Within the corridor of 20 km agricultural land constituted 66.7 per cent, while forests comprised 25.5 per cent. The KR alignment did not pose any threat to tall evergreen forest, major formation of mangrove or wild life sanctuaries.

According to statistics published by Director of Agriculture, forest constituted 31.24 per cent of the region's area prior to construction of KR in 1989-90. This per cent increased to 33 after its completion in 1997-98. According to Forest Survey of India forest area increased from 46.8 to 50.42 per cent during this period. The forest cover increased in 15 tahsils, while it declined in 10 tahsils through which KR is constructed. Area under cultivation and land not available for cultivation has declined over the years in tahsils through which KR is constructed. Wasteland comprises 18.5 per cent of the region's total area. Land with or without scrub occupies 63 per cent of total wasteland. NAS comprises one-fourth of the region's total area, which has increased during this period except in Raigad district. Fallow land has increased in all the districts. The impact of KR on industrial development is visible in the analysis of land use changes on 1 : 50,000 scale for the region around Roha. New industrial area has come up along the alignment after the construction of Konkan railway. To conclude, construction of Konkan Railway has improved the environment and land use of the region.

REFERENCES

Bhat, D.M., (October, 2001). Forest dynamics in tropical rain forests of Uttar Kannada district. in Western Ghats, *Current Science*, Vol. 79, No. 7.

Bilsborrow, R. E. & Pamela DeLargy, (1991). Land use, migration and natural resource deterioration: The experience of Guatemala and the Sudan, in Kingsley Davis, (ed) *Resources, Environment and Population, Present Knowledge, Future Options*, Oxford University Press, New York.

Developing Maharashtra's Wastelands, (June 1993). Directorate General of Information and Public Relations, Government of Maharashtra, Mumbai

Environmental Impact Assessment for the section of Konkan railway line in Goa State, Vol. I, Report, March 1992, RITES, New Delhi

Environmental Impact Assessment Studies for West Coast (Konkan) Railway line Corridor Survey using satellite (IRS-LISS-II) data, Report submitted to RITES. National Remote Sensing Agency, Hyderabad (year not mentioned)

India, Policies and Issues in Forest Sector Development, Report No. 10965-IN, (June 24, 1993) Document of the World Bank, South Asia Regional Office.

Jha, C.S., Dutt and Bawa, Deforestation and Land use changes in Western Ghats, India, *Current Science*, Vol.79, No. 2, 25 July 2000. pp. 231-237.

Pilot project on natural resource accounting in Goa, (Phase 1) Final Report. Prepared for Directorate of Planning and Statistics, Goa. TERI Project Report No. 99, RD 61, New Delhi.

Time bound programme for development of Konkan (1997-2000). Brochure published by Directorate of Information and Publicity, Government of Maharashtra, Mumbai.

4

Population Growth: Rural and Urban

Population growth leads to accelerate the demand for infrastructure. The experience world over reveals that with the advent of railways obscure villages have grown into towns and several towns into big cities. It has been observed that commissioning of Konkan railway has led to higher growth of population which is manifested in urbanization of the region. The impact of population growth is reflected mainly in growth of existing towns and emergence of new towns. The construction of Konkan railway commenced in 1990. The 1981 and 1991 census provide the necessary data base for comparative analysis of population growth prior to construction of Konkan railway, and 2001 census provides the latest statistics of rural and urban population for our study region after the commissioning of Konkan railway. A comparative analysis of growth rates of urban and rural population for the past two decades is attempted below. Population growth for the two consecutive decades (1991-2001and 1981-91) is calculated for stations on Konkan railway, towns newly emerged on Konkan railway and for towns of 1991 census located on railways and National Highways, as they reflect the impact of infrastructure development. Changes in the composition of urban population by size class of towns are analysed. Indian census has been presenting the demographic data separately for rural and urban areas. The unit of classification in this regard is "town" for urban areas and "village" for rural areas. Village level statistics required for this study are compiled from District Census Handbooks and Primary Census Abstracts of respective districts. The issue of population displacement is discussed first in the context of Konkan Railway.

Population Displacement: Changing Connotations in the Indian Context

The urgency for development of infrastructure for overall economic development of the nation is accompanied by increased resistance and opposition to land acquisition and the resultant involuntary displacement. The issue of population displacement had been of varying importance in our country in different phases of development after Independence. In the first phase of "primacy of economic growth and development" up to 1960s, population displacement was considered as a secondary issue. Population displacement became an emerging issue only during the second phase (1970s and 1980s) of "Development with social justice and a human face". In the Third phase (the 1990s) of "Development under structural adjustment programmes" population displacement became a mandatory concern. With the advent of the New Economic Policy, population displacement has been increasing at a rate not experienced earlier. Urban, industrial and port development projects undertaken in this phase have sometimes led to multiple displacement of affected populations. While data on investments and projects are available, information about the land acquisition and displacement by these projects is difficult to obtain. Until recently development induced displacement of population was considered a "sacrifice" some people have to make for the larger good. However, perceptions are changing because of delays in project implementation and benefits foregone; growing awareness about the potential adverse economic, social and environmental consequences of population displacement; and increasing concern about people's welfare. For the development policy makers, displacement is a social cost you pay for the 'greater common good.' But for those who pay this social cost, it is alienation from home and hearth, a forced eviction out of one's occupation and one's cultural milieu. It is a process that invariably leads from 'spatial-economic security' to 'spatial-economic captivity'. The repercussions of development induced population displacement are often bitter.

Population Displacement by Development Projects

A project set up in any area displaces not only those who hold land in their possession, but equally also those who are co dependents on the system, including tenants, share croppers, landless labourers and also those carrying on any trade, occupation, etc. All such persons are affected and deprived of their livelihood. Displacement results from the following developmental processes: urbanization and industrialization, natural resource development, agricultural transformation and infrastructure expansion. It is extremely difficult to provide any new infrastructure in India without displacing individual business or residences. The high population density in India means that it is difficult to acquire land without displacing people in large numbers. Population displacement can be both direct and indirect. Direct displacement is a coerced eviction from one's home. Indirect displacement occurs when development projects lead to ecological deterioration and therefore force people to move. Many development projects that require involuntary displacement of people generally have adverse economic, social and environmental impacts on the displaced people. Project affected persons stand to lose, as a consequence of the project, all or part of their physical and non-physical assets, including homes, communities, productive lands, resources such as forests, pasture lands, fishing areas or important cultural sites, commercial properties, tenancy, income earning opportunities, social and cultural networks and activities. Displaced people are the ones occupationally displaced by development projects with no alternative avenues of employment The affected persons see acquisition as not only leading to physical dislocation from their land and loss of income but as also causing social, cultural and psychological problems in relocation in new and different surroundings. Field surveys have revealed that a large number of people have lost their land, their jobs and their income due to development projects in India. Ongoing World Bank projects in Asia are estimated to displace over 1.5 million people. An example of the magnitude of population displacement in Bank-assisted projects is the Jingjiu Railway Project (1994) in the People's Republic of China which involves the displacement of about 210,000 people. However, numbers alone may not present a full picture of the intensity of impact on the local people.

Displacement due to Konkan Railway Project

On this background we have to study the displacement and resettlement in connection with the Konkan railway—a project of transport infrastructure expansion. Long alignments required for roads or railways cause resettlement effects along a narrow right of way. How many in the Konkan have been displaced in the name of this infrastructure development? How many of the displaced have found an alternative niche? What is the quantum of social cost we have paid for this infrastructure development? Is the 'greater common good' bought by this social cost equitably distributed? These are the disturbing questions which are difficult to answer. Konkan railway passes through 3 tahsils of Raigad and Ratnagiri districts each, 5 tahsils of Sindhudurg district, 4 tahsils of North Goa and 3 tahsils of South Goa district. The magnitude of displacement and land acquisition for Konkan railway project can be visualized by the number of villages affected tahsilwise as shown below in Table 4.1. Similar statistics for coastal Karnataka region were not available.

Table 4.1: Land Acquisition and Displacement in Maharashtra and Goa

Tahsil/ District	*No. of villages affected*	*Total*	*Area acquired in hectare*	*No. of persons affected*	*Persons affected as % of Tahsil population*	*Area aquired as %. of Tahsil area*
Ratnagiri District						
Khed	35	215	339	1913	1.12	33.6
Chiplun	16	163	206	1674	0.79	17.1
Sangames.	16	194	208	2071	1.00	16.6
Ratnagiri	17	199	1002	2389	1.22	13.1
Lanja	20	122	310	2496	2.35	0.40
Rajapur	12	216	318	1531	0.90	26.5
Raigad District						
Roha	9	170	71	311.	0.22	10.0
Mangaon	33	244	156	1182	0.64	16.7
Mahad	21	183	200	1124	0.60	2.6
Sindhudurg District						
Sawantwadi	09	140	155	1679	0.90	11.5
Kudal	17	123	143	2153	1.5	17.6
Malwan	03	135	24	410	0.35	3.9
Kankavli	17	105	288	3096	2.40	37.4
Vaibhavwadi	09	58	98	1169	2.40	23.5
North Goa	19		316	1584	0.23	0.18
South Goa	34		546	2056	0.41	0.28

Source: (1) Konkan Railway Corporation Ltd., Regional Office, Ratnagiri (Maharashtra).

(2) Office of Special Land Acquisition Officer, Madgao (Goa).

The ratio of number of persons affected per ha of land acquired was calculated for all the tahsils of Maharashtra through which Konkan railway passes. The ratio ranges from the lowest of 2.38 persons in Ratnagiri tahsil to the highest of 17.08 persons in Malwan tahsil of Sindhudurg district. Over 10 persons per ha were affected in all the tahsils of Sindhudurg district. Percentage of villages affected to total number of villages in a tahsil ranges from the maximum of 16.3 in Khed tahsil of Ratnagiri district to the minimum of 2.2 per cent in Malwan tahsil.

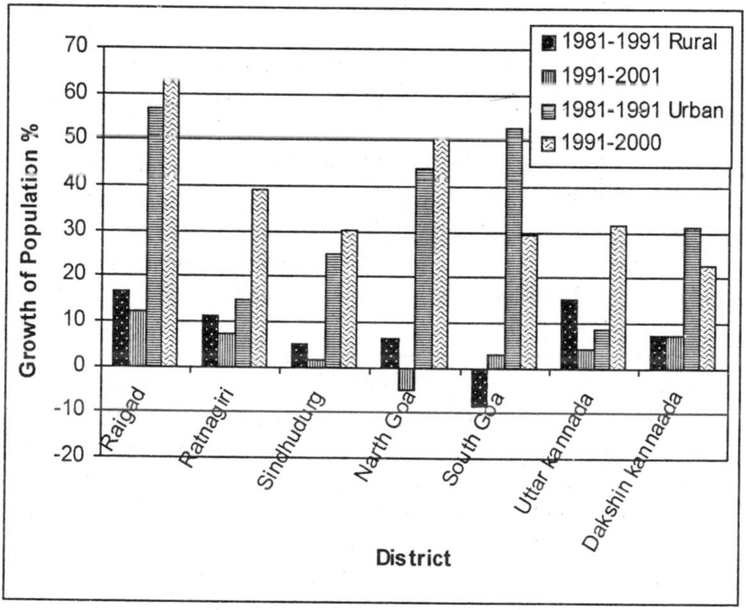

Fig. 4.1: Rural Urban Variatisms in population growth

Distribution and Growth of Population

As per the latest 2001 Census of India the total population of our study region exceeds one crore (10,467,320). It constitutes 1.02 per cent of India's population. The population size at district level ranges from the highest of over 30 lakhs in Dakshina Kannada (DK) to the lowest of 5.8 lakhs in South Goa district. Population density in our study region (243 persons) is much lower than the all India average of 324 in 2001. The pressure of population in the tiny state of Goa is very high compared to other coastal areas.

Maximum population density of 436 persons per sq. km is recorded in North Goa district followed by Dakshina Kannada (416). These two districts record population density higher than the national average. The lowest density of 132 persons is recorded in Uttar Kannada district. In terms of rural population as well, Dakshina Kannada is the most populous district, followed by Raigad and Ratnagiri districts. Table 4.2 gives the population density and increase in population pressure. Population density in our study region increased from an average of 215 persons per sq. km in 1991 to 243 in 2001 with an average addition of 28 persons per sq. km. This average addition is lower than the national average of 57 persons. Both Raigad and North Goa districts have the highest addition of persons per sq. km (53) in spite of their varying decadal growth rates. The lowest addition of 5 persons per sq. km is recorded in Sindhudurg district due to its lowest growth rate. The districtwise decadal growth rates for rural and urban population for the last two decades are shown in Table 4.2.

Table 4.2: Population Density and Growth

(growth rate, per cent)

District	*Growth rate, per cent*				*Population Density Persons per km²*	
	Rural population		*Urban population*			
	1981-91	*1991-01*	*1981-91*	*1991-01*	*1991*	*2001*
Raigad	16.79	12.09	56.59	62.73	255	308
Ratnagiri	11.28	7.35	14.69	39.19	188	207
Sindhudurg	5.10	1.52	24.95	30.48	160	165
North Goa	6.57	-5.01	43.99	50.79	383	436
South Goa	-8.35	2.83	52.77	29.29	257	298
Uttar Kannada	15.23	4.48	8.73	31.72	119	132
Dakshina Kannada	7.52	7.31	31.12	22.65	319	356
Average region	7.73	4.37	30.78	35.55	215	243

During the decade 1991-2001 the population in this region increased by 12.98 per cent This growth rate is far lower than the Indian average of 23.44 per cent. Rural population increased by 6.4 per cent which is much below the national average of 18.3 per cent The table reveals that decadal growth rates of rural population have declined in the subsequent decade, whereas growth rates of urban population have increased. The growth rates of rural population are far lower than those of urban population. This is very well illustrated in Fig.4.1. The slow rural growth and decline of rural population suggests out migration, i.e. movement of people towards urban centers within or outside the region, which is facilitated by infrastructure development. The highest decadal population growth of total population during 1991-2001 is recorded in Raigad district (21.56%) and the lowest in Sindhudurg district (3.72%). The difference in rural growth rates of two decades is the lowest in DK district while it is the maximum in UK district. In other words, rural population growth rates of the two decades do not differ much in DK district, while UK district recorded a much lower rural growth in 1991-2001 compared to 1981-1991.

Population Growth around Konkan Railway

In order to study the impact of infrastructure development on rural population growth, growth rates of rural population are calculated for the region around Konkan railway. For this purpose, village level and tahsil level data are utilized, and the tahsils through which Konkan railway is constructed are taken into consideration. Villagewise population growth is calculated for the villages whose land was acquired for KR construction and for the stations on Konkan railway. The stations on Konkan railway are grouped as villages and towns or urban centres. 30 stations are classed as rural and 21 as urban (towns). They are shown in the maps showing population growth by size class of towns (Figs. 4.10 to 4.13). Population growth during 1991-2001 is calculated for all the stations and summarized in Table 4.3.

Table 4.3: Population growth of stations on Konkan Railway

Population size of villages	*No. of stations*	*Population Growth % (Rural)*	*Size class of towns*	*No of stations*	*Population Growth % (urban)*
Less than 1000	2	21.56	Class I	2	43.45
1000-2000	11	5.64	Class II	3	23.28
2000-3000	8	9.61	Class III	7	17.99
3000-5000	3	7.52	Class IV	7	16.36
5000-10000	3	7.83	Class V	2	9.98
Over 10000	3	11.65	—	—	—
Total	30	—	—	21	—

The highest urban growth of 44.85 per cent is recorded at a class I town Udupi, whereas the lowest urban growth of 4.45 per cent is recorded at Goregaon, a class V town in Raigad district. It can be generalized from Table 4.3 that as the size class of towns increases, its population growth also increases. There is no clear cut trend of population growth variations with population size of villages. However, the largest and the smallest size classes have recorded higher growth rates. The population of 5 villages (stations) has declined. Four out of them belong to population size of 1000—2000. The highest rural growth of 57.6 per cent is recorded at Mangaon in Raigad dist. belonging to the same size class. One village records a marginally increase in population which can be termed as rural stagnation. The impetus provided by Konkan railway to urbanization is, to some extent, at the expense of rural population decline.

Rural Population Growth

Rural population growth is now studied at tahsil level for the tahsils through which KR is constructed and at village level for the affected villages whose land was acquired by KRCL for construction work. The following table shows population growth of affected villages, along with total villages in tahsils of Raigad district.

Table 4.4: Population growth of villages located on Konkan Railway

Tahsils through which KR is constructed	*Number of villages whose land was acquired*	*Number of villages where population declined*	*Population of affected villages, 1991*	*Population of affected villages, 2001*	*Per cent increase in rural population of*	
					Affected villages	*Total villages in the tahsil*
Roha	9	2	7550	8600	14.0	2.84
Mangaon	30	6	27775	36084	29.9	-21.45
Mahad	21	8	17768	17979	1.2	6.70

Source: Compiled from the data collected from the Corporate office of Konkan Railway and Census of India, 1991 and 2001.

Konkan Railway passes through the above 3 tahsils of Raigad district. Among them, Mangaon tahsil experienced overall decline in its rural population, while the other two tahsils reveal rural stagnation. The population of 16 villages of these tahsils has declined. 42 out of the 60 affected villages are small in population size, having less than 1000 population each. But it is remarkable that the villages located on Konkan railway in Roha and Mangaon tahsils, whose land was acquired have grown much faster in population compared to rest of the villages. High growth rate of affected villages in Mangaon can be attributed to emergence of a new town Utekhol, which was the largest among the affected villages (1991 population 4351). It recorded an increase of 67.5 per cent and is designated as a town in 2001 census. Two other large size villages, Chinchwali and Lonere (having population of 1000 to 2000 in 1991) also recorded extraordinarily rapid growth. This is because they attracted development on account of their locational advantage. A Technological Institute for higher education with specialized courses was established at Lonere by Government of Maharashtra. Large villages on Konkan railway in Mahad tahsil have failed to attract such development and hence the population of villages like Karanjadi and Dasgaon (2000 to 3000 population in 1991) has either declined or grown at a slow pace. A major factor responsible for this is the scarcity of water in this region.

In Ratnagiri district, among the 6 tahsils through which Konkan railway is constructed, the situation in four tahsils can be described as rural stagnation (rural growth 2 to 6%) and one tahsil (Lanja) recorded decline in rural population. Only Ratnagiri tahsil recorded a remarkable rural growth of 14.12 per cent In Sindhudurg district, out of the four tahsils penetrated by Konkan railway, rural population recorded negative growth (decline) in two tahsils and low rural growth (2 to 5%) in Kudal and Sawantwadi tahsils. In North Goa district, among the three tahsils, only Tiswadi recorded an increase in rural population by 7.13 per cent and rural decline in two others. In South Goa district, rural population declined in Salcete tahsil, recorded a slow growth (2 to 5%) in two tahsils and the highest growth of 14.3 per cent in Quepem. In Uttar Kannada district rural population declined in Karwar and Ankola tahsils and recorded a moderate rural growth of about 10 per cent in two others. In Dakshina Kannada district Mangalore tahsil recorded the highest rural growth of 22 per cent followed by Coondapur (19.34%). These are the highest rural growth rates amongst all the tahsils penetrated by Konkan railway. Udupi tahsil recorded a low growth of 7 per cent. From the above analysis it is evident that the rural areas along the Western coast are characterized by stagnation or out migration and the impact of infrastructure development on rural growth is restricted only to three tahsils—Ratnagiri, Coondapur and Mangalore, and a few large size villages.

Growth of Rural Population by Size Class of Villages

The pattern of rural settlements varies considerably in the sub -regions of this coastal belt. Differential growth of rural population around Konkan railway has an impact on the settlement pattern in these regions. This impact can be seen in Table 4.5.

Table 4.5: Changes in No. of Villages by Size Class

Size class of villages	*Maharashtra*		*Karnataka*	
	1991	*2001*	*1991*	*2001*
Less than 200	9	9	—	—
200-500	65	58	—	1
500-1000	64	65	6	4
1000-2000	26	33	16	20
2000-5000	4	3	43	50
5000-10,000	1	1	12	12

Source: Compiled from Villagewise Primary Census Abstract of 1991 and 2001 census.

In this table Roha tahsil of Raigad district represents the settlement pattern in Maharashtra and Mangalore tahsil represents Karnataka. There is predominance of small size villages in Maharashtra, (majority with less than 1000 population), whereas the villages in Karnataka are of fairly large population size (over 2000 population). Number of villages in higher size class has further increased during the decade in Karnataka. In Maharashtra, number of villages in one size class has increased, while there is a decline in small size villages. This is due to overall low growth rate of rural population in this sub-region.

Occupational Structure of Workers

A town or a city is distinguished from a rural settlement both by its function and size. It is the non-agricultural function which is basic to the development and growth of towns. Their large size is the consequence of a widening functional base. By and large, most towns started as non-agricultural settlements and grew into larger and more complex cities. Purely agricultural settlements grew into local service centers. The statistics of occupational structure of workers published in the census help us to explain the variations in rural and urban growth rates. The break up of workers by nine industrial categories for each district in 1991 census is taken into consideration. The important villages and towns in the Konkan region are concerned mainly with commerce and transport. It is interesting to note that the dependence of workers on agriculture is significantly higher in coastal Maharashtra (73%) compared with other areas. It is the lowest (24%) in Goa. The proportion of workers in manufacturing is the maximum (28.13%) in DK district and is extremely low (3 to 6%) in Ratnagiri, Sindhudurg and UK districts. The proportion of workers in trade and commerce and other services is relatively higher in Goa compared with other districts. Other services record the maximum of 36.5 per cent of workers in the urban population of North Goa due to the presence of State capital Panaji and due to prevalence of tourism related services. Tourism has become a major industry with a number of starred hotels along the beaches and in towns. The proportion of workers in transport, storage and communication in South Goa district (10.85%) is the highest in the region due to the presence of Mormugao port and Margao railway junction. In spite of the predominance of manufacturing activities, DK district records slow urban growth, and in spite of the prevalence of large agricultural workforce, Raigad district records the highest rural and urban growth in the region. Hence, it is difficult to establish any clear-cut relationship between the functional base and population growth at district level.

Sex Ratio

A special feature of the population of the western coast is the sex ratio, i.e. the number of females per 1000 male population. Average Indian sex ratio (933 females per 1000 males) reveals predominance of male population over female population. However, our study region has the predominance of female population. The sex ratio of our study region is 1028. This indicates a very large out migration of male population from these coastal districts to the metropolitan city of Bombay (Mumbai), to other parts of India and also abroad in Gulf countries in search of employment in industries and other establishments. All the districts of our study region record sex ratio higher than the national average. The highest sex ratio of 1135 is recorded in Ratnagiri district and the lowest of 951 in North Goa district. Ratnagiri district holds first rank in sex ratio in Maharashtra.

Male dominated out migration for better employment opportunities is the main cause for poor growth rate and increasing predominance of females in the region. Urban sex ratio reveals a different picture. Generally rural areas have a higher proportion of females than the urban areas. Hence urban sex ratio is always lower than total sex ratio. The deficiency of females in urban areas is considered to be due to in-migration of male working members. In the urbanization process single male workers migrate to urban areas in search of employment opportunities leaving their families behind in rural areas. At all India level the average urban sex ratio is 901 whereas in our study region it is 956. Raigad district records the lowest urban sex ratio of 884 while DK district records the highest urban sex ratio of 1013 in 2001.

Changes in Sex Ratio

During the decade 1991-2001 India's average sex ratio has gone down from 940 in 1991 to 933 in 2001. Within our study region the sex ratio declined from 1047 in 1991 to 1028 in 2001. In 5 districts the sex ratio in 2001 is lower than in 1991. Raigad district had an excess of females over the male population in 1991, while in 2001 the sex ratio of 975 is recorded. Even in Ratnagiri and Sindhudurg districts the large preponderance of female population has relatively gone down. In South Goa and Uttar Kannada districts there is a marginal increase in sex ratio in 2001 over the 1991 census. As compared to significant decline in sex ratio of total population urban sex ratio in our study region has declined marginally from 959 to 956. In three districts—North Goa, Uttar Kannada and Dakshina Kannada the urban sex ratio increased during the decade.

Towns and Urban Agglomerations in the Region

In Census of India, 2001 two types of towns are identified:

(a) *Statutory towns*: All places with a municipality, corporation, Cantonment board or notified town area committee, etc. so declared by a state law.

(b) *Census towns*: Places which satisfy the following criteria:

   (i) a minimum population of 5,000;

(ii) at least 75 per cent of male working population engaged in non-agricultural pursuits; and

(iii) a density of population of at least 400 persons per sq. km. Towns with population 100,000 and above are termed as cities.

Urban agglomeration is a continuous urban spread constituting a town and its adjoining urban outgrowths (OGs), or two or more physical contiguous towns together and any adjoining urban outgrowths of such towns. Examples of outgrowth are railway colonies, university campuses, port area, military camps etc. that may have come up near a statutory town or city. For Census of India, 2001, it was decided that the core town or at least one of the constituent towns of an urban agglomeration should necessarily be a statutory town and the total population of all the constituents should not be less than 20,000 (as per 1991 Census). With these two basic criteria having been met, the following are the possible different situations in which urban agglomerations could be constituted.

(i) a city or town with one or more contiguous outgrowths;

(ii) two or more adjoining towns with or without their outgrowths; and

(iii) a city and one or more adjoining towns with their outgrowths all of which form a continuous spread.

There are nine urban agglomerations in our study region in 2001. The combined population (11.12 lakhs) of UA's constitutes 39.71 per cent of the region's urban population. In 1991 there were six UA's in the region with a combined population of 8.23 lakhs. They constituted 40.79 per cent of the region's urban population in 1991. Four UA's were added in Uttar Kannada district—Karwar, Bhatkal, Kumta, and Ankola. With the inclusion of a new town Venkatapura, Bhatkal in Uttar Kannada district assumed the status of UA. Margaon, (South Goa) which was treated as UA in1991 is not treated as such in 2001. One of its constituent town Aquem is treated as separate towns in 2001 census and another one—Raia has been declassified. This has reduced the share of UAs in the urban population of 2001 compared to 1991. There are no urban agglomerations in the coastal districts of Maharashtra. Five Urban Agglomerations—Mangalore, Udupi, Margao, Mormugao and Karwar are located on railway line.

The distribution of towns and Urban Agglomerations in the region is as follows:

Table 4.6: Towns and Urban Agglomerations

District	*Statutory towns*	*Census towns*	*Total*	*Urban Agglomerations*
Raigad	11	15	26	—
Ratnagiri	4	6	10	—
Sindhudurg	3	2	5	—
North Goa	7	20	27	1
South Goa	7	10	17	1
Uttar Kannada	11	2	13	5
Dakshina Kannada	12	14	26	2

Spatial Distribution of Towns in Konkan

A large number of trading ports dominated the Arabian sea very early in history. Chaul, Uran, Nagothana, Roha, Shrivardhan, Bankot, Jaigad, Kharepatnam and Malwan are some of the very old port towns of Konkan. Inadequacy of the inland transport often led to the development of coastal towns on the sea creeks far inland, to command a larger area. In fact, sea creeks commanded the best sites for the sailing vessels which did not require much draught and could easily ply a considerable distance inland, giving rise to small break of bulk towns. Some of these port towns existed as early as the first century and are mentioned by many ancient voyagers. Irrespective of size, it appears that the towns were more numerous on the coast and received the attention of foreign traders. Water transport was the most economical and safest mode of transport in the early days and a creek-head location was both a site for handling maritime trade as well as a central place occupying a focal point in the collection and distribution of regional goods. These towns have not grown much and the advent of steam navigation has made them defunct as international ports. They are now no more than small fishing centres with country crafts plying in and out. Most of the towns were linked with transport routes, particularly the routes that traversed the Western Ghats from the Konkan to the plateau. They have a remarkable similarity in their geographical location and economic functions. On the coastal side, each estuary is a focus of human activity, mostly agricultural, and partly marine. With the construction of National

Highway from Mumbai with Kanyakumari (NH 17) a large number of places along the western coast were connected, increasing the accessibility of towns and leading to their rapid population growth. This NH was the forerunner to the development of Konkan railway. It has been observed that large villages function as market/service centres in respect of several surrounding villages and gradually grow into small towns and attain urban status in the course of time. This is applicable to our study region as well. Weekly market centers in our region are getting strengthened with infrastructure development. In the Indian scenario there is a tendency for corridor/ribbon development along the railway line in the vicinity of villages and towns. Unplanned development occurs near the railway stations and flyovers by appearance of fruit marts, pan shops, *dhabas*, etc. Emergence of new towns along Konkan railway is discussed separately.

The important villages and towns located along the Konkan railway in Raigad district from north to south direction are Roha, Kolad, Indapur, Mangaon, Goregaon, Veer and Karanjadi. In Ratnagiri district towns on the railway line are Khed, Chiplun, Sangameshwar, Ratnagiri and Rajapur. The towns on the alignment in Sindhudurg district are Kharepatan, Kankavli, Oras, Kudal. Sawantwadi and Banda are slightly away (about 10 km) from the railway line. Important stations in Goa are Pernem, Thivim, Karmali, Verna, Madgaon, Bali, Canacona. Stations along the route in Uttar Kannada are Karwar, Ankola, Gokarna, Kumta, Honavar, Murdeshwar and Bhatkal. In Dakshina Kannada district the towns are Byndoor, Coondapur, Gurpur, Barkur, Udupi, Mulki, Surathkal, Thokur and Mangalore. Their locations along Konkan railway and in relation to NH 17 are shown in Figs. 10 to 13.

Level of Urbanization in the Region

The demographic distribution pattern between the urban and rural areas is undergoing a major change all over the world with people moving from rural areas and conclaves to urban areas and urban fringes. Urbanization is the most important social transformation of our times. It is the epitome of a society advancing towards industrial and technological enrichment. Urbanization refers to the proportion of the total population concentrated in urban settlements or an increase in the proportion of urban population to total population over a period of time.

Table 4.7: Level of Urbanization

(per cent of urban to total population)

District	*1991*	*2001*
Raigad	17.83	24.24
Ratnagiri	8.96	11.32
Sindhudurg	7.60	9.56
North Goa	33.94	44.92
South Goa	50.33	56.02
Uttar Kannada	24.17	28.67
Dakshina Kannada	28.31	31.10
Average of 7 districts	22.30	26.76

The level of urbanization in our study region always remained lower than the all India average. According to the Census of 2001 there are 2,800,902 urban inhabitants constituting 26.76 per cent of the region's total population. For all India this proportion is 27.78 per cent The state of Goa is highly urbanized with 56 per cent urban population in South Goa district. Coastal Karnataka has a comparatively higher level of urbanization than Coastal Maharashtra. DK district has a higher level of urbanization than UK district. Sindhudurg district of Maharashtra records the lowest level of urbanization (9.56%) in the region. In 1991 with 2,066,233 urban inhabitants, urban population constituted 22.30 per cent of total population of the region, against the all India average of 25.72 per cent. Only 3 districts of our study region had higher level of urbanization than the national average in 1991, whereas in 2001 this number increased to 4. However, the ranks of districts in terms of level of urbanization have remained the same in both the census years.

Urban Growth

Transport is the basic infrastructure or a prerequisite in the process of urbanization. The scale and rate of urban growth in India has led to serious deficiencies in the availability of basic urban infrastructure and the delivery of essential urban services. Wherever the government provided massive infrastructure in terms of provision of road transport, electrification, water supply, expansion of administrative, educational and health facilities and establishment of industrial estates, all of them gave a spurt to urbanization process.

The impact of infrastructure development is reflected in the process of urbanization of this region. It is remarkable that average urban growth has accelerated in this region, i.e. urban growth rate in 1991-2001 (35.55%) is higher than in 1981-91 (32.42%), and is higher than the national average of 31.39 per cent. All the districts recorded a much higher urban growth during 1991-2001 as compared with 1981-91. DK district is the only exception where urban growth rate recorded in 1991-2001 (22.63%) is lower than that in 1981-91 (31.12%). Therefore, though the level of urbanization has increased all over the region, it has declined in this district. The difference in urban growth rates during the two decades is more pronounced in case of Ratnagiri and Uttar Kannada districts. Here the process of urbanization accelerated at a higher rate compared to previous decade. Urban growth in Raigad district has accelerated after the emergence of Konkan railway, from 54.23 per cent in 1981-91 to 65.2 per cent in 1991-2001, which is the highest in the region (Fig. 4.1). This is associated with the highest rural growth (12.1%) and the lowest urban sex ratio in the region (884). Major factor for such enormous growth is migration and job opportunities. The influence of Bombay (Mumbai) metropolitan city was so tremendous that it prevented the development of any other class I city in nearby districts. It is only after the planned efforts of CIDCO and Government of Maharashtra to develop New Bombay (Navi Mumbai) as counter magnet to saturated Bombay Metropolitan region that urbanization received momentum in Raigad district. Therefore, the nucleus of population growth in the north is around Navi Mumbai and Panvel. Ratnagiri and Sindhudurg districts, the homeland of Konkan culture, is one of the least urbanized tract in the region (per cent of urban to total population 10%). The decadal growth rate of total population over here is strikingly low (less than 10%. Urbanisation in Goa is associated with rural depopulation. DK district is conspicuous in the region for its slow urban growth (22.65%). Population growth recorded by most of the towns during 1991-2001 is much lower than that in 1981-91. The addition of urban population during the decade is not so high as compared to rural addition.

Addition of Rural and Urban Population

The actual addition or increase of urban population in this region during 1991-2001 (7.39 lakhs) is much more than that of rural population (4.62 lakhs). The addition of both urban and rural population is the highest in Raigad district, followed by DK district.

Urban addition is the lowest in Sindhudurg district. The addition of urban population is more than rural population in all the districts except Ratnagiri. (Fig.4.2). In this district more rural inhabitants have been added (over one lakh) compared to urban inhabitants (54,093) because of its low level of urbanization. It is evident that infrastructure development has accelerated the growth of both rural and urban population in this district. Goa experienced decline in rural population by 14,912 persons during the decade. The rural population of North Goa has declined and the addition of rural population during the last decade in South Goa district is the lowest in the region. Rural population decline in Goa is attributed to falling birth and death rates. The birth rate in Goa declined from 19.5 in 1991 to 17.85 in 2000. The death rate declined from 7.5 to 7.16 per cent and infant mortality rate from 18.7 to 14 (*Source*: Directorate of Planning, Statistics and Evaluation, Government of Goa). The decline in the above statistics in Goa can be attributed to higher literacy and income levels, a higher proportion of working women, higher age at marriage, good response to pulse polio programme and improved health care for women (Goa Report, 2001). The report also points out that Goa has reached replacement level in population growth rate.

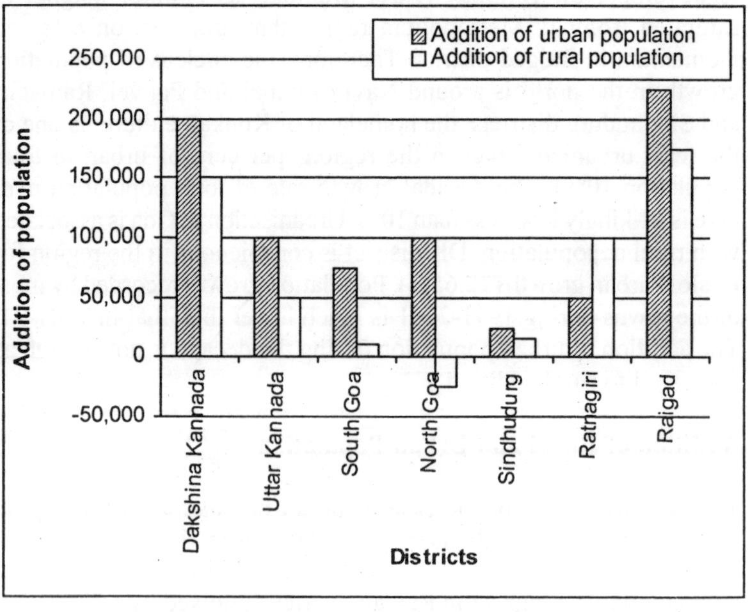

Fig. 4.2: Addition of Rural and Urban population, 1991-2001.

Distribution of Towns by Size Class

A special feature of urbanization in this region is the small size of urban centres. Among the 124 towns as per 2001census, 94 towns have the population size of less than 20,000 each, and seven among them are of less than 5000 population. A continuous stretch starting from Ratnagiri to Uttar Kannada district is devoid of class I cities. One class I city (Panvel) has developed in the extreme north (in Raigad district) and two cities (Mangalore and Udupi) are located in the extreme south in Dakshina Kannada district. Over one-fifth (23%) of the region's urban population lives in these three cities. In Raigad district Panvel emerged as a class I city in 2001 accommodating 19.52 per cent of the district's urban population. There are 2 class II towns (50,000 to 99,999 population) in this district—Khopoli and Navi Mumbai. They contain 26.38 per cent of district's urban population. Ratnagiri, the only class II town of Ratnagiri district, contains 41 per cent of the district's urban population. Sindhudurg district has no class I or class II towns. In North Goa district, capital Panaji (class II city) contains 20.5 per cent of the district's urban population which point out that the urban scenario of Goa is dominated by small towns. In South Goa district, Mormugao urban agglomeration accounts for 31.85 per cent of district's urban population. Mormugao and Madgaon, two class II cities contain 53.4 per cent of district's urban population. In Uttar Kannada district half (49.9%) of the district's urban population lives in 3 class II towns—Dandeli, Karwar and Sirsi. In Dakshin Kannada district two class I cities—Mangalore and Udupi accommodate 58 per cent of the district's urban population. There is no class II town in this district.

Changes in Size Class of Towns

The process of urbanization is reflected an increase in average population size of a town in the region (urban population/no. of towns) from 20,060 in 1991 to 22,588 in 2001. Of the 103 towns of 1991 census 54 retained the same size class, 21 were upgraded to higher size class and 2 assumed a lower size class in 2001 census. From a single class I city of Mangalore in 1991 the number of class I cities in the region increased to three in 2001with the addition of two cities, Panvel (Raigad) and Udupi (Dakshina Kannada). Consequently, the share of class I cities in the region's urban population increased from

13.68 per cent in 1991 to 23.08 per cent in 2001. As the number of class II towns declined from 11 in 1991 to 9 in 2001, the share of urban population in these towns declined from 31 per cent to 23.2 per cent. With the increase in number of class III towns from 12 to 18, the share of urban population in these towns increased from 17.3 to 20 per cent. Due to emergence of new towns the number of class IV towns increased from 34 to 44 and class V towns from 35 to 43. However, the proportion of urban population in both the classes declined. Due to extremely low share of urban population in class VI towns, the decline in their number is of little consequence. With varying growth rates of existing towns, the composition of urban population by size class in each district has undergone changes as illustrated in the diagrams (Figs. 4.3 to 4.9).

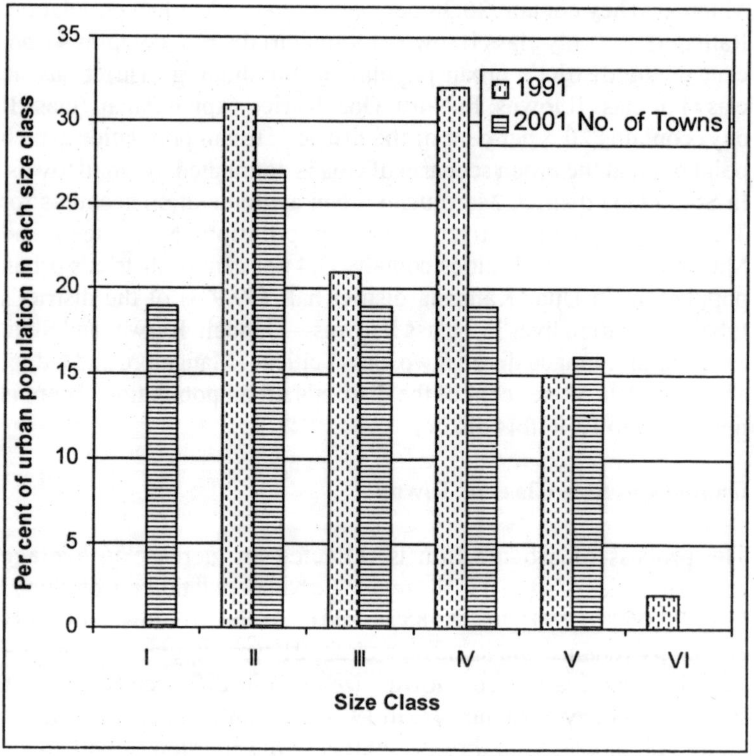

Fig. 4.3: Changes in Composition of Urban Population by Size Class Raigarh District

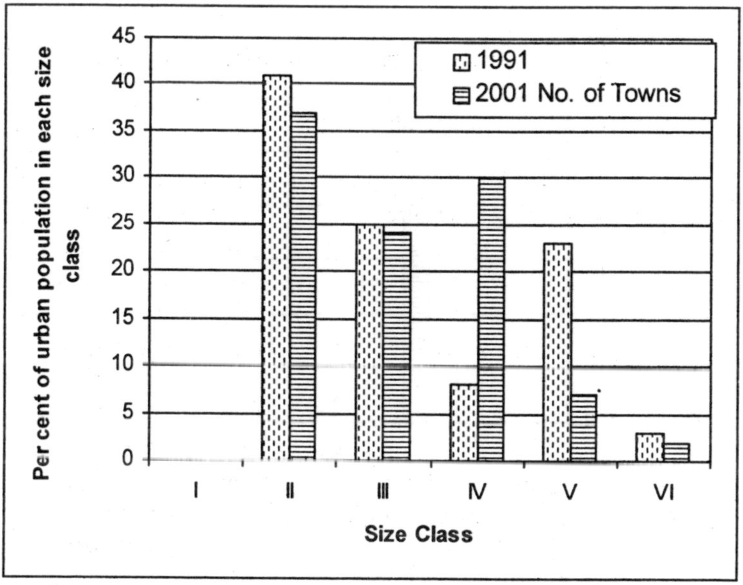

Fig. 4.4: Changes in Composition of Urban Population by Size Class Ratnagiri District

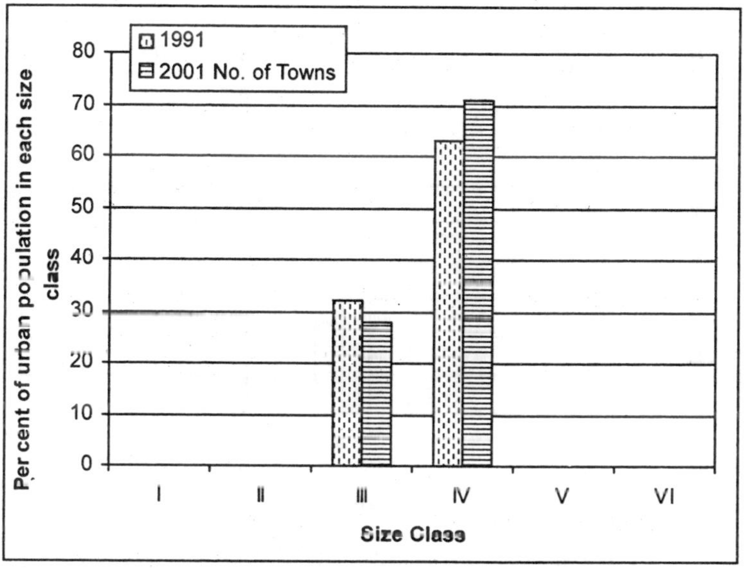

Fig. 4.5: Changes in Composition of Urban Population by Size Class Sindhudurg District

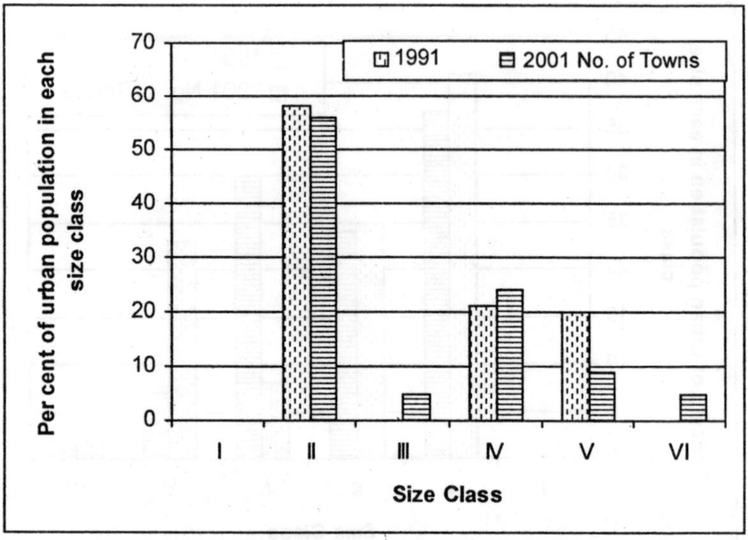

Fig. 4.6: Changes in Composition of Urban Population by Size Class South Goa District

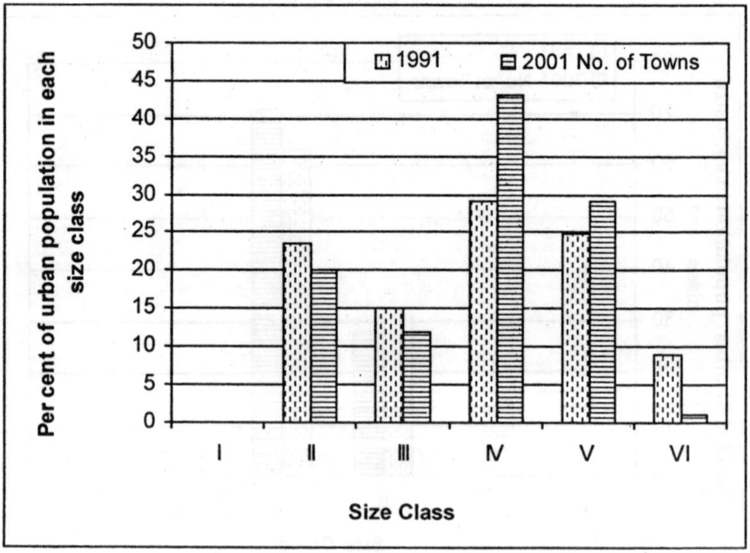

Fig. 4.7: Changes in Composition of Urban Population by Size Class North Goa District

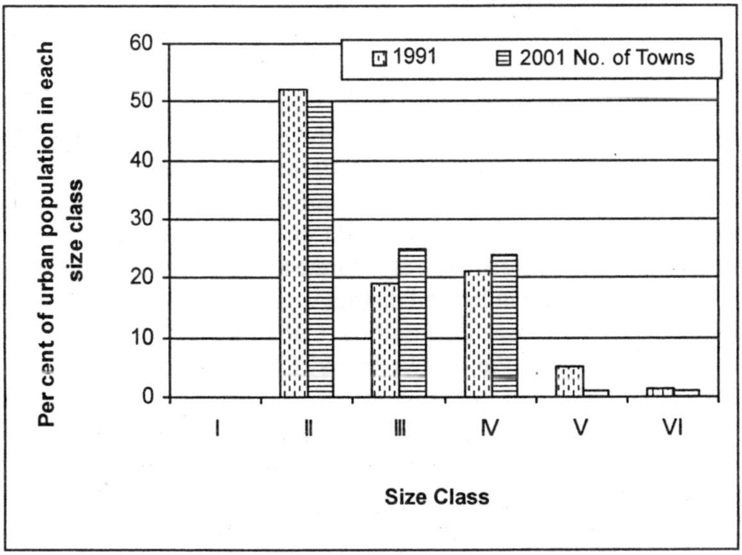

Fig. 4.8: Changes in Composition of Urban Population by Size Class Uttar Kannada District

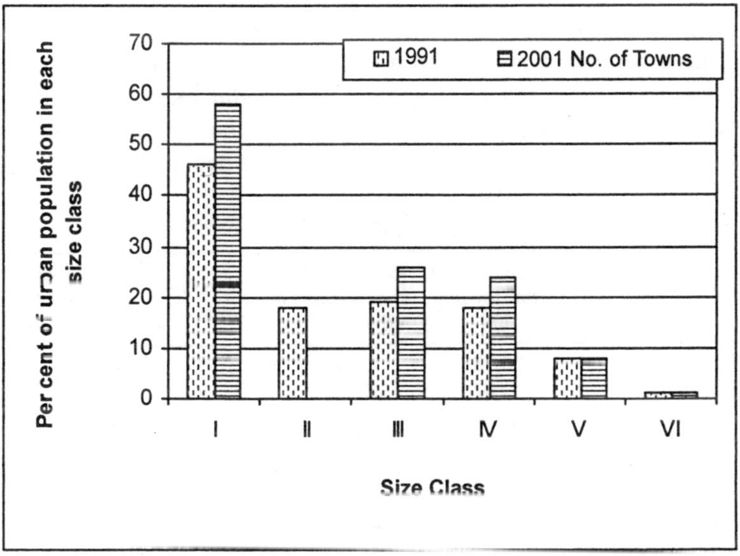

Fig. 4.9: Changes in Compositon of Urban Population by Size Class Dakshin Kannada District

Declassified/Deleted/Merged Towns

In all 23 towns in the region were declassified, deleted or merged with nearby towns and lost their urban status during the decade 1991-2001. Among them are five class IV towns and 4 class V towns. The largest number of towns (11) was deleted in Dakshina Kannada district. Three places designated as project colonies in 1991, viz. Magod and Kadra in Uttar Kannada and Hosangadi in Dakshina Kannada district cease to exist as separate towns in 2001. Three towns in Ratnagiri and one in Raigad district were declassified. No town in Sindhudurg district was declassified. Cortalim near Mormugao and two constituents of Panaji UA in 1991—Tadaigao and Jua were deleted in 2001.Two constituents of Mangalore U A in 1991—Surathkal and New Mangalore port were merged with Mangalore and cease to exist as separate towns in 2001. Surathkal (1991 population 55,925) is the largest town declassified/deleted. Three constituent towns of Udupi UA in 1991—Malpe, Badgabettu and Puttur were merged. Kemminje was merged with Puttur and Peramunnur with Ullal. All the declassified or deleted towns except Surathkal and Baindur in DK were "census towns" and not statutory towns. The combined population of deleted towns as per 1991 census was 102,961. In the process of addition and deletion, North Goa seems to have benefited most with a net increase of 9 towns in 2001 compared to 1991. In Uttar Kannada and Dakshina Kannada districts the number of deleted towns is more than the newly emerged towns and the total number of towns in these districts declined by one each.

Table 4.8: Emergence of "New" Towns

District	*Population of new towns, 2001*	*Per cent share of new towns in urban addition*	*No. of new towns, 2001*	*No. of deleted Towns, 2001*
Raigad	58,169	27.56	8	1
Ratnagiri	40,870	75.55	5	3
Sindhudurg	14,725	76.44	1	0
North Goa	77,616	67.71	11	2
South Goa	49,556	66.53	6	2
Uttar Kannada	36,188	38.73	3	4
Dakshina Kannada	76,640	44.38	10	11
Average 7 districts	—	47.83	—	—
Total	*353,764*		*44*	*23*

Emergence of "New Towns"

A special feature of urbanization on the Western Coast is the emergence of new towns. The number of towns in the region increased from 103 in 1991 to 124 in the 2001 census. Apparently, there is an increase in number of towns by 21 during the decade. However, a detailed district level analysis reveals that this is an outcome of 44 newly emerged towns minus 23 deleted or merged towns. The location of towns in each district as per 1991 and 2001 census years are marked in Fig. 4.10 to 4.13. In the 2001 census 44 places which satisfied the demographic criteria were treated as "census towns". Of the 44 new towns included in the urban frame,13 belong to class IV, 25 class V and 6 class VI towns. Among the new towns Sulya in Dakshina Kannada district is the most populous (2001 population—18,026), whereas Guhagar in Ratnagiri district is the smallest in population size (population 3205). The maximum number of new towns is observed in North Goa (11) followed by Dakshina Kannada district (10). Only one town has been added in Sindhudurg district. Many of the new towns are parts of Urban Agglomerations, e.g. six new towns are located around Mangalore UA, and one each close to Panaji and Udupi are treated as separate towns.

New Towns Located on Konkan Railway

Emergence of new towns in this region having low level of urbanisation assumes significance, as they have developed with infrastructure development, i.e. along the newly commissioned Konkan railway. After the commissioning of Konkan railway, seventeen places (villages) located on the Konkan railway are designated as towns for the first time in the 2001 census—four in Raigad district, two in Ratnagiri, one in Sindhudurg, three in Goa, one in Uttar Kannada and six in Dakshina Kannada districts. In addition to this, six new towns are located on South Central Railway passing through this region—four in South Goa and two in Dakshina Kannada district. The population of new towns located on Konkan railway (1.64 lakhs) constitutes 40.75 per cent of the population of all the new towns. The emergence of new towns on Konkan railway reflect the impact of infrastructure development on urbanization. The locations of new towns in Maharashtra, Goa and Karnataka by size class are shown in Figs. 4.10 to 4.13.

Population growth of "New Towns"

The emergence of new towns has made a substantial contribution to urbanization of the region. Decadal growth rates are calculated for all the 'new towns'. In order to study the impact of Konkan railway on the new towns and population growth of new towns *vis-a-vis* the existing towns is compared. All the new towns in 2001 census were treated as villages in the 1991 census. Their population in 1991 is compiled from the Village Directory of 1991 District Census Handbooks of respective districts. The new towns reveal tremendous variations in population growth ranging from over fourfold increase to decline in population. In general, because of their small size, new towns have recorded average growth rate (30.63%) lower than the region's average urban growth rate (35.5%) 14 out of 44 new towns recorded growth rate higher than the region's average urban growth rate. Two of them recorded exceptionally high growth rates. The population of three new towns declined. The population of new towns (3.5 lakhs) constitutes 47.83 per cent of the actual addition of urban population during the decade. The sex ratio of new towns (953) is lower than the average urban sex ratio (956).

New Towns along Coastal Maharashtra

Eight new towns have been added in Raigad district, 7 of them are class V. Four of them are located on Konkan railway and the rest on or in the vicinity of National Highway. Ambivali south of Panvel is the only case of a tiny village (1991 population 302) which emerged as a class V town in 2001 (population 6796). Birwadi and Utekhol towns close to Konkan railway and NH recorded extremely high growth rates of 62 and 68 per cent which may be attributed to the establishment of industries in their vicinity. Nagothane and Mohpada recorded moderate growth of 35 to 46 per cent All these towns have a very low sex ratio of less than 850 which is due to male selective in migration. Dhatav, in spite of its low growth (6.44%) records a low sex ratio of 789. The contribution of new towns in the urban addition (27.56%) is the lowest in Raigad district. This is due to the small size of new towns and the higher growth rate of existing large towns.

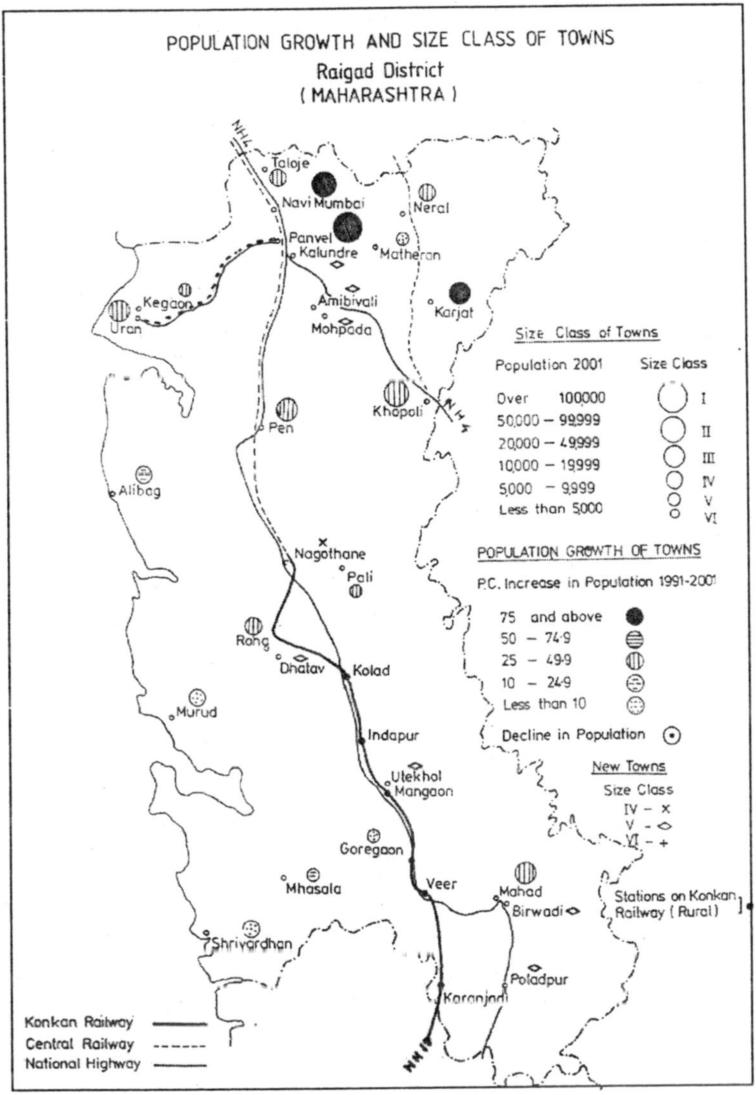

Fig. 4.10: Population Growth and Size Class of Towns, Raigad District

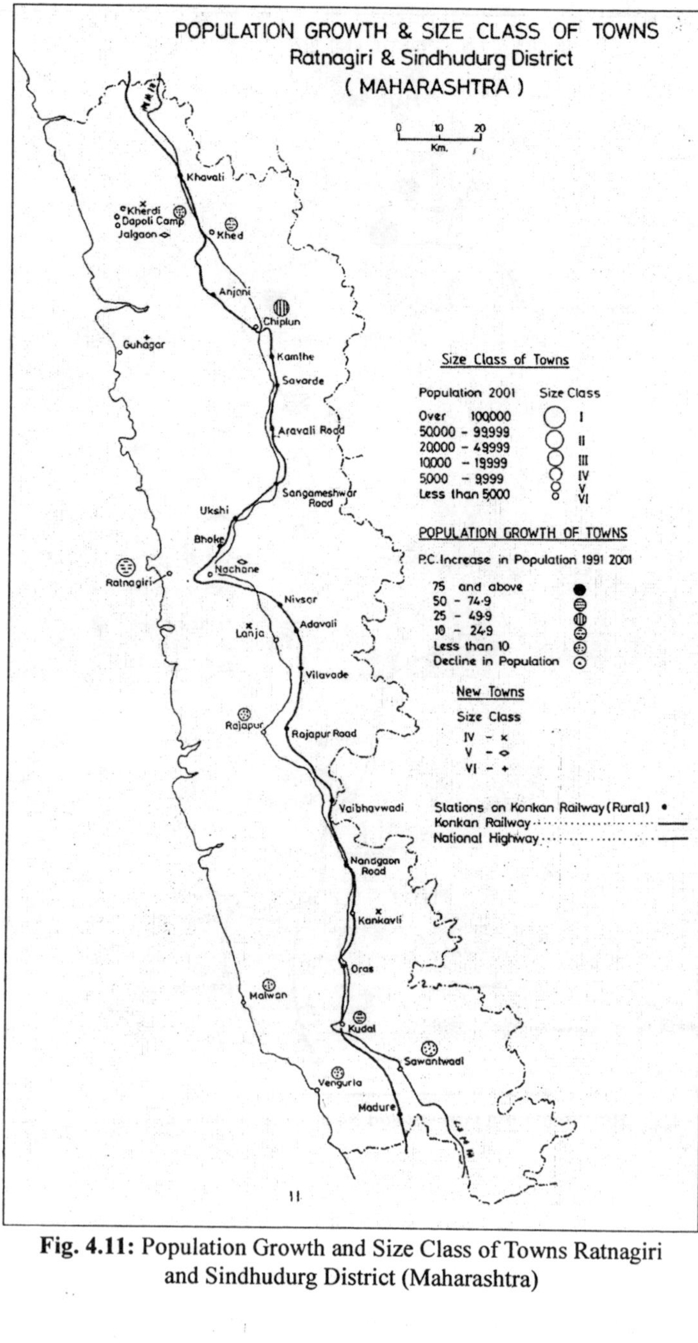

Fig. 4.11: Population Growth and Size Class of Towns Ratnagiri and Sindhudurg District (Maharashtra)

Lanja, a tahsil headquarter in Ratnagiri district located on NH 17 emerged as a new town by recording a high growth of 40 per cent. It was a village of over 5000 population in 1991. Nachane, another new town near Ratnagiri located in between Konkan railway and NH also recorded high growth rate (53.5%). Kherdi near Dapoli camp emerged as a new class IV town by recording over fourfold increase because of the establishment of an industrial estate by the State Government. This is the only town in the district which records a low sex ratio of 889, confirming the male selective in migration towards the industrial establishments, while rest of the towns record a high sex ratio of over 940. Guhagar, a tiny coastal town is an example of a new town which has experienced decline in population in 2001 as compared to 1991 census. Kankavli, a tahsil headquarter located on Konkan railway and NH 17 is the single new town in Sindhudurg district. In spite of its location and size (class IV town in 1991) it has recorded a moderate growth of 24.8 per cent which is the highest among all the existing towns of the district. This district is characterized by extremely low level and rate of urbanization in the region. In the above districts new towns (5 and 1 respectively) have contributed to over three-fourth of their urban addition during the decade (75.55 per cent and 76.44 per cent respectively).

Goa

Among the new towns of North Goa Aldona, Socarro, Goa Velha are located on or close to Konkan Railway and Bandora, Calapur, Chimbel,. Colvale, Curti, Quela, Saligao are located on or close to National Highway. The growth rate of new towns in Goa varies from an extremely low rate of 2.54 per cent at Parcem located away from main transport lines to the highest of 90 per cent at Davorlim on South Central railway close to Margao UA. In spite of the location on Konkan railway Goa Velha is a stagnant town (growth rate 7.6%) whereas Socorro to the north recorded a high growth of 60 per cent. Among the new towns of Goa the population of Aldona (North Goa) and Aquem (South Goa) has declined. Aldona has a very high sex ratio of 1155. Chicalim, in spite of its extremely low sex ratio (730) records a low growth of 7.6 per cent In South Goa district San Coale, a rural component of Mormugao UA in 1991 is treated as a separate new town in 2001. Because of its industrial base and its location on NH 17 A, it recorded a high growth of 58.4 per cent It also records a low sex ratio of 840. Curti close to NH 4A and San Jose de Areal on

South Central railway also belong to the category of high growth rate. Curti records a low sex ratio (874), while San Jose de Areal has a high sex ratio (950). Other new towns in Goa have recorded slow to moderate growth up to 30 per cent. Thus the association between high sex ratio and low growth of towns cannot be clearly established in case of Goa. Expansion of tourist infrastructure and tourism based industries have led to emergence of new towns and higher growth of certain towns in Goa. Due to the largest number of new towns (11) North Goa district recorded a higher urban growth rate (50.8%) than South Goa district (29.3%), However, the contribution of new towns in district's urban addition in North Goa (67.71%) and South Goa district (66.5) is more or less the same.

Coastal Karnataka

Being a forested and underdeveloped region, the rate of urbanization in UK district is low (31.7%). The main towns in the area are situated in the coastal plains. Though 3 new towns emerged in 2001 census, 4 towns of 1991 census were declassified. Hence the contribution of new towns in the district's urban addition is low (39%). Among the three new towns, Siddapur and Mundgod were listed as towns in the 1981 census but were declassified in the 1991 census. Both are located in the interior. Mundgod in spite of being away from NH or railway line recorded a higher growth (54%) and lower sex ratio (942). Venkatpura, a part of Bhatkal UA located on Konkan railway recorded a high growth of 45.44 per cent It has a very high sex ratio of 1063.

In Dakshina Kannada district 10 new towns emerged, whereas 11 towns lost their urban status, deleted or merged. New towns contribute 44 per cent of the district's urban addition. The highest growth recorded by a new town in this district is 40.7 per cent by Mudushedde, which is a constituent of Mangalore UA. New towns like Beltangdi and Sulya were listed as towns in 1981 census and deleted in 1991. All the new towns except two (Thokur and Sulya) have a predominance of female population, i.e. sex ratio exceeding 1000. It reaches the maximum of 1172 at Mulur, and the lowest of 922 at Sulya. Thus low urban growth is associated with high sex ratio. In spite of the location on Konkan railway or NH new towns have recorded slow to moderate growth. As Mangalore was already located on Southern railway and connected with the Indian railway system, the commissioning of Konkan railway has not made a significant impact on population growth of new towns located on it.

Growth of towns located along Konkan Railway

Population growth for two consecutive decades 1981-91 and 1991-2001 is calculated separately for towns located on railways and National Highways passing through this region. Prior to commissioning of Konkan railway, this region was partly accessed by Central railway in Raigad district, South Central railway in Goa and Southern railway in DK district. Apart from National Highway No. 17 which runs almost parallel to Konkan railway and cuts across all the 7 districts, NH 17A connects Mormugao port to NH 17 and Konkan railway. NH 4 going to Belgaum and Bangalore passes through Raigad district and by NH 4A passing through Goa, Panaji is connected to it. A total of 33 towns are located on Konkan railway. In addition to these towns 25 'new towns' (treated as towns for the first time in 2001 census) are located on railway lines. Out of them 18 are located on or very close to Konkan railway. All the towns located on railway lines and National Highways have recorded higher urban growth during 1991-2001, i.e. after the commissioning of Konkan railway compared to previous decade (1981-1991). The average population growth rates by size class for the towns located on railways and National Highways passing through this region are as follows:

Table 4.9: Growth of towns located on railways and National Highways

Size class	*No. of towns*	*Average growth rate*
I	03	54.58
II	07	36.42
III	13	23.00
IV	15	19.65
V	06	12.81

In general, it is observed that population growth rate declines as the size class of towns decreases. Among the towns located on railway lines, 9 towns recorded growth rates higher than the region's average of 35.55 per cent. They include all the class I towns, viz. Panvel, Udupi and Mangalore, one town belonging to class II, viz. Navi Mumbai, three towns belonging to class III—Pen and Karjat (Raigad) and Ullal (Dakshina Kannada), Taloje (Raigad) and Navelim (South Goa) are two class IV towns. The functional base of these towns, i.e. composition of main workers in 9 major industrial activities as published in the 1991. Census is taken into consideration while explaining the variations in their population growth.

Coastal Maharashtra

Panvel an industrial township close to Navi Mumbai is the largest town in Raigad district. It emerged as a class I city for the first time in the 2001 census by registering a very high urban growth of 76.85 per cent. In the previous decade (1981-1991) the growth rate was 58.73 per cent. An important junction of National Highway and Central Railway, Panvel serves as a link between Mumbai and Konkan railway. A branch of railway line also connects Panvel to Uran and newly developed Nhava Sheva port renamed as Jawaharlal Nehru Port Trust (JNPT). It is a fully modernized port. A waste heat recovery plant of 120 MW is installed at Uran. Thus Panvel serves as a link between the two major ports. The extension of suburban train services up to Navi Mumbai (Belapur) and new road access via Thane creek bridge up to Panvel further accelerated its population growth to emerge as a class I city. The construction of the Thana creek bridge has resulted in bringing the area close to Greater Bombay. A new airport project for Navi Mumbai is also proposed at Panvel. Its functional base is diversified, with the maximum proportion of workers in manufacturing (29.3%) followed by other services (23.6%) and trade and commerce (22.5%). One tenth of its workforce is engaged in transport and related activities. CIDCO shifted the wholesale market of iron and steel located at Mumbai to Kalamboli in Panvel tahsil at a planned site of over 325 ha. Rasayani, a separate industrial township close to Panvel (1981 census population 3651) was declassified in 1991 and merged with Panvel.

Navi Mumbai is a planned township and recorded the highest (91.68%) population growth among all class II towns of our study region. It is associated with the lowest sex ratio of 771 among all the towns in the region. It possesses all the basic modern infrastructure facilities for rapid urban growth in the future. If this city is excluded from the list of class II towns, the average population growth of remaining 4 towns is reduced to 24.34 per cent. Navi Mumbai is basically an industrial town with about half (46.4%) of total workforce engaged in manufacturing activities, 17 per cent in primary activities, followed by other services (12.9%) and transport (8.1%). The two major industrial zones of Navi Mumbai are the Thane Belapur Industrial Area in the West and the Taloje industrial estate in the east.

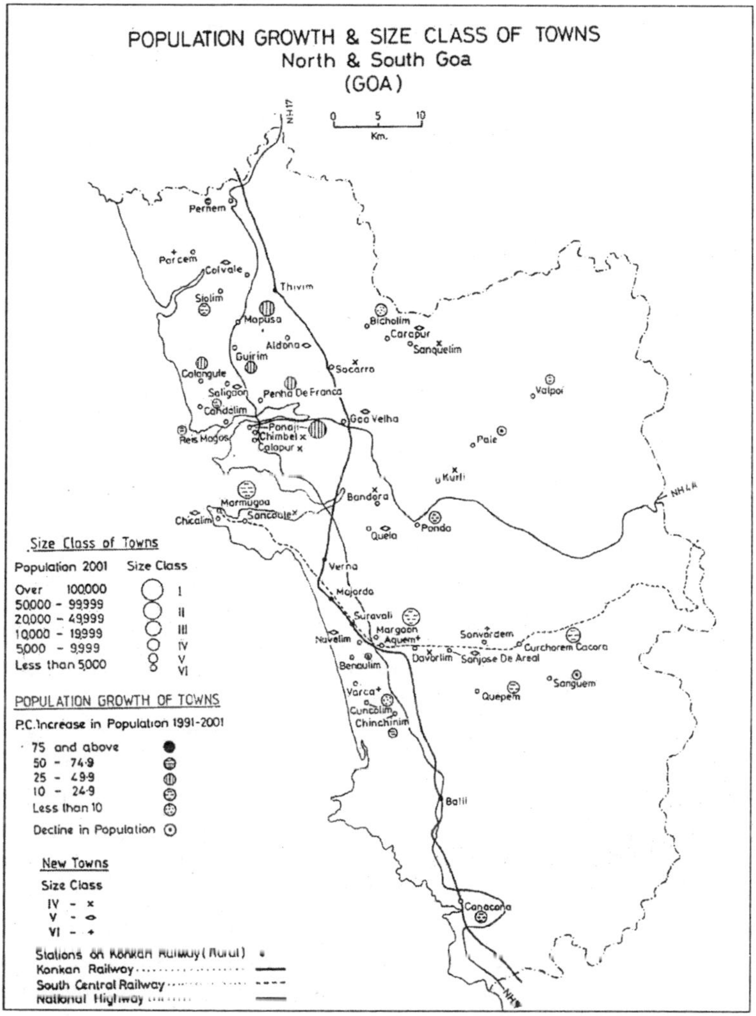

Fig. 4.12: Population Growth and Size Class of Towns North and South Goa

Two small towns in Raigad district (Pen—class III, Taloje—class IV) recorded high population growth ranging from 40 to 45 per cent because of their locational advantage on railway and national Highway (Fig 4.10). Pen is developed as a service centre, trade and transport town. It is famous in Maharashtra for manufacturing of Ganesh idols. Taloje, close to Navi Mumbai is an industrial and commercial centre having a low sex ratio (839). Karjat recorded over three-fold increase in its population and from a class V town in 1991 became a class III town in 2001 because of its locational advantage. It is the only town in the region which has entered next higher size class by such an extraordinarily rapid growth rate. It is located on Central railway and National Highway 4 connecting Mumbai to Pune. An industrial estate is located nearby which provided an impetus for its rapid growth. Roha, in spite of being on Konkan railway and a centre of manufacturing (engaging 36% of workforce) recorded moderate population growth (29.2%) and a relatively higher sex ratio (932).

The process of urbanization has accelerated in Ratnagiri and Sindhudurg districts after the construction of Konkan railway, as the towns located on it—Khed, Chiplun, Ratnagiri recorded higher growth rates after the emergence of Konkan railway (1991-2001) compared to previous decade (1981-1991) (Fig. 4.11). Chiplun (class III) recorded the highest urban growth rate of 34.7 per cent. An industrial estate located at Lote Parshuram near Chiplun by Maharashtra Government attracted several chemical industrial units and accelerated its population growth. Chiplun is a trade and service centre (engaging 24.5% of total workers in each activity), while Ratnagiri is a centre of diversified activities with service as the dominant one. Ratnagiri, (class II) the district headquarters, lags behind the smaller towns in population growth (24.5%). Chiplun records a higher sex ratio (981) compared to Ratnagiri (898). Enron Corporation of USA completed a major power project at Dabhol in this district. But being away from Ratnagiri town, it has yet to make an impact. Kudal (class IV) in Sindhudurg district recorded moderate growth rate of 17.7 per cent and a high sex ratio of 972. It is a centre of household industry engaging 19.2 per cent of total workforce. This proportion is the highest among all the towns located on Konkan railway. Service, trade and manufacturing are the other activities worth noting in this small census town. Other towns in the district are almost stagnant (recording less than 10% of growth). Even the rural population growth is negligible (1.52%).

Goa

Panaji, the capital of Goa is not located on railway line, but at the intersection of NH 17 and 4A. NH 4A further crosses Konkan railway. Due to Goa's increasing importance as a popular tourist destination after the commissioning of Konkan railway, and increasing flow of foreign direct investment in tourism sector, Panaji recorded the highest urban growth (35.64%) amongst all the towns of Goa during 1991-2000, whereas during 1981-1991 it recorded a very slow growth (less than one per cent). Mormugao UA has emerged as a class I UA in 2001 census. However, Mormugao Municipal Corporation is a class II town, and is more populous (97,000) than neighbouring Margao (78,000) with whom it is connected by South Central railway. Its connectivity has markedly improved after the commissioning of Konkan railway. Vasco, its nearest rail head is not yet treated as a town. Mormugao port on its integration by Konkan railway has been opened up to a much larger hinterland and connected with rest of the country. The fast growth in the port trading activities resulted into a heavy immigration of labour in this town during the eighties. Further the large scale construction work taken up on the port and in the town also necessitated import of labour. This resulted in a lower sex ratio (884). The town has now become an important trading centre with growing warehousing and storage facilities. The proportion of workers engaged in transport and storage (26.5%) and other services is equally large (26.9%). The port has stimulated expansion transport related jobs. The other development that has taken place around the town is the establishment of a fertilizer factory at San Coale and a number of ship and barge building workshops.

A power project was completed at San Coale in July 1999. The main reason for Mormugao's rather slow growth (16.45%) during this decade compared to previous decade (19.41%) is the emergence of San Coale as a separate new town in 2001 census. Mapusa on Konkan railway recorded relatively higher growth (27%) compared to Ponda (20.6%) a temple town and a pilgrim centre (Fig 4.12).

Margao/Madgao is the headquarter of South Goa district and an important junction of Konkan railway, South Central railway and National Highway. It has become one of the few railway stations in the country to achieve the prestigious ISO 9001 : 2000 certification for providing better amenities and services to ticket holders. It experienced accelerated urbanization after the completion of Konkan railway, i.e. recorded higher urban growth than the previous decade Its growth rate during 1991-2001 (33%) is next to Panaji. The main economic activity of the town is trade and commerce (30.4%) followed by service (28.3%) manufacturing (13.9%), construction and transport. It is known for its industries manufacturing various items like beverages, chemicals, food products, etc. Navelim (class IV), a constituent of Margao UA in 1991

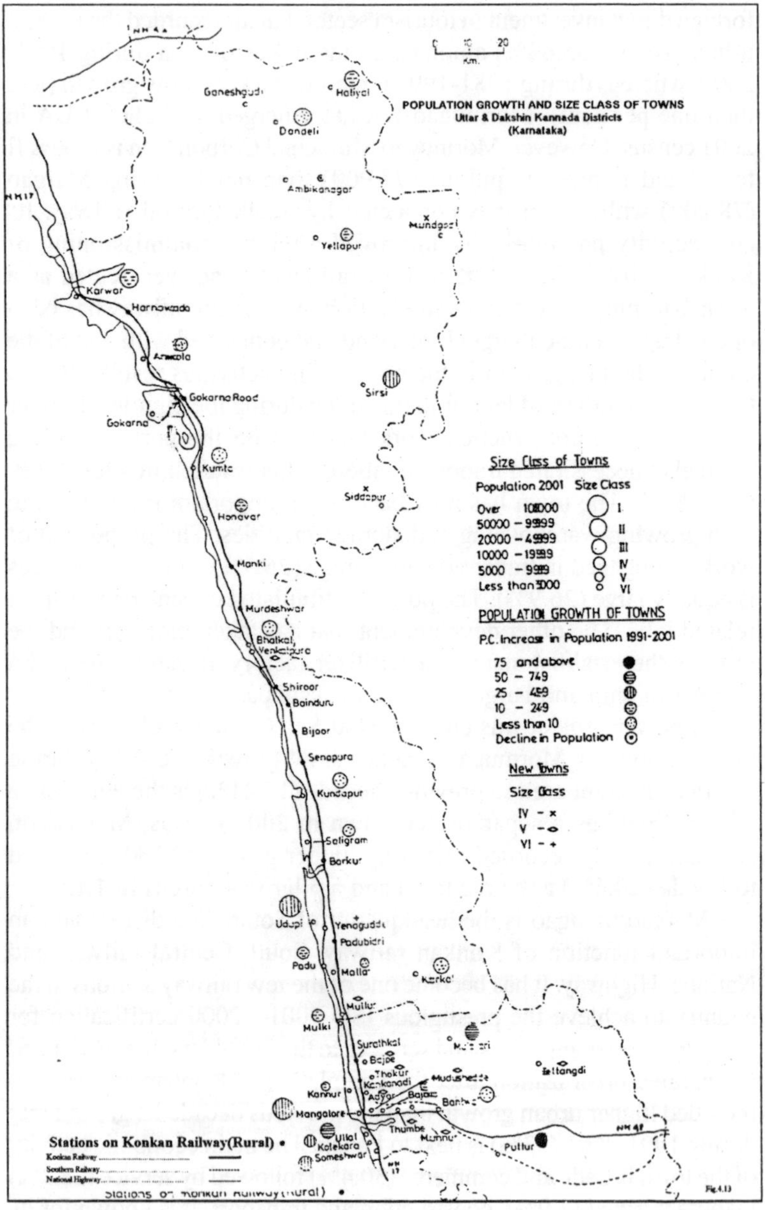

Fig. 4.13: Population Growth and Size Class of Towns Uttar and Dakshina Kannada Districts (Karnataka)

is treated as a separate town in 2001 census. It recorded a high growth of 41 per cent and a lower sex ratio of 918 compared to Margaon (968). This can be attributed to the commissioning of Konkan railway. The maximum proportion of workers in this town is engaged in trade and commerce (24.3%) followed by services and manufacturing activities. Other towns on Konkan railway in Goa—Cuncolim, Canacona and Pernem belonging to class IV and V recorded population growth of 10 to 15 per cent. They are all service and trading centres. Cuncolim records a very high sex ratio of 1049.

Coastal Karnataka

Most of the class IV and class V towns in Uttar Kannada district in spite of their location on Konkan railway recorded extremely low population growth ranging from virtual stagnation in Bhatkal and Kundapura (growth rate 1 and 0.4%) to the maximum of 16 per cent. Their growth rates are even lower than in the previous decade. This trend is in harmony with rural population growth, which is also lower than the previous decade and is much lower than urban population growth. The main reason behind their slow growth is their excessive dependence on primary activities (over 35% of total workers in them) and low level of industrial activity. They mainly serve as centres of commerce and are unable to attract population. Karwar is treated as UA for the first time in 2001 census. The population of Karwar Municipal Council increased by 23.43 per cent but if the population of Chitakula, its urban outgrowth is taken into consideration, the UA recorded growth of 47 per cent. Accelerated urbanization is noticed in case of Karwar and Sirsi towns. They had witnessed a very low growth during the previous decade. Karwar port was once an important trade centre of the British, French, Portugese and the Arabs. Lack of connectivity by broad gauge railway line with the hinterland in Karnataka hampered its development. Modern Karwar is mainly a service and trading centre (% of workers in other services 31.4, trade and commerce 21.3, sex ratio 930). Manufacturing activities are yet to pick up (9.5%) and dependence of workers on primary activities is substantial (16.5%). Karwar is being developed as a naval base of Indian Navy under Operation Sea Bird project. According to Karnataka Chamber of Commerce, there is a very large potential for exports through Karwar for the various products of North Karnataka. Kodasalli hydel power project on Kali river was completed in August 1999. A coke oven cum regeneration power plant based on imported coal with an investment of Rs. 3300 mn is proposed. This is expected to boost urbanization in the future.

Only three towns Puttur, Ullal and Mulbidri in Dakshina Kannada district recorded substantially higher urban growth rates in 1991-2001 (60 to 86%) compared to previous decade (17 to 28%). Puttur is located on Southern railway and Ullal is a constituent of Mangalore UA located on southern railway and National Highway 17 (Fig. 4.13). Puttur records a relatively lower sex ratio of 983 compared to Ullal (1032). Their high growth is mainly due to merger of adjoining towns. Kemminje, a 1991 census town was merged with Puttur in 2001. Similarly, Peramunnur was merged with Ullal. The Karnataka state identified the coastal region, with Mangalore constituting the primary urban centre as a priority area for investment. Mangalore was commissioned as the ninth major port of the country in 1974. It was further expanded as an export port for iron ore from Kudremukh Co. With the establishment of oil refinery, petrochemicals and fertilizer plant and numerous ancillary units, Mangalore emerged as a leading industrial town. With these developments an atmosphere leading to the overall development of the entire district was generated. An LPG bottling plant was completed in February 2000. A new LNG terminal is also to be commissioned. Manufacturing is the dominant activity (engaging 36.1% of workers). In addition, an equally high proportion (21%) of workers is engaged in trade and commerce and other services. Transport and communication activities engage another 9.3 per cent of workforce. In spite of industrial development, Mangalore and Udupi UA recorded a much lower population than projected in their Environmental Master Plan. Mangalore UA (Sex ratio 991) recorded a lower growth of 26.49 per cent compared to previous decade (39.11%) as two of its constituent towns were declassified. Someshwar, a non-municipal census town (class III) near Mangalore, recorded a high urban growth of 31.4 per cent and a higher sex ratio of over 1017. Kotekara, a part of Mangalore UA, in spite of its small size (class IV) recorded a high growth of 47.5 per cent and a high sex ratio of 1099.

Udupi was carved out as a separate district from Dakshina Kannada district. Though upgraded in status as district headquarters, population growth of Udupi UA (7.91%) is much lower than the growth of Udupi CMC (44.85%). Three towns around Udupi in 1991 were declassified in 2001 census. Udupi recorded a very high growth of 133 per cent during 1981-91, which declined to 44.85 per cent in 1991-2001. An important station on Konkan railway, Udupi is famous for its temples and cuisines. It is developed mainly as a commercial center with the largest proportion of workers (36.3%) engaged in trade and commerce. In fact Udupi has the highest proportion of workers in this activity among all the towns on Konkan railway. Next in importance are other services (28.9%) and manufacturing (17.2%).

Growth of other Towns Along the Coast

When the towns along the railway lines and National Highways are excluded, the growth rate of remaining towns throughout the region is extremely low. As the Konkan railway is constructed away from the coastline in the coastal districts of Maharashtra, all the small towns located along the coast have remained virtually stagnant. Even in Raigad district, which recorded the highest urban growth of 62.7 per cent, population of small coastal towns like Shrivardhan and Murud increased by just 3 per cent. Only Alibag, the district headquarters recorded moderate growth of 20 per cent because of its diversified functional base. Coastal towns as well as other interior towns in Ratnagiri and Sindhudurg districts are also stagnant. Sindhudurg district records the lowest growth of total population in the region. The case of towns in Goa is somewhat different. Goa is a highly urbanized tiny state with predominance of small towns. North Goa recorded higher urban growth (50.8%) than its neighbouring districts. Konkan railway and National Highway in Goa have a coastal alignment and coastal towns are more dynamic because of their scenic beauty and natural resources. Therefore the towns located along the Konkan railway and other coastal towns do not differ much as regards their decadal growth rates. Even class IV and class V towns recorded growth of over 15 per cent Coastal Karnataka recorded urban growth lower than the region's average. A few class IV and V towns along the coast recorded a low urban growth of less than 10 per cent while other towns in the interior Uttar Kannada recorded moderate growth of 15 to 25 per cent. Mulbidri town in Dakshina Kannada district recorded a very high growth of 62 per cent in spite of its interior location while nearby Karkal, a tourist place and known for Kudremukh iron ore project recorded an extremely low growth of 4 per cent. If the medium and minor ports in coastal Karnataka are developed, perhaps this may lead to some increase in urban population. The slow growth of small towns over here suggests that commissioning of Konkan railway might have helped in out migration or outward movement of population from these towns, due to their increased accessibility with rest of the country. Excessive predominance of females in the population of these towns also support this generalization.

Conclusion

Urbanisation on the Western coastal region gained momentum as a consequence of infrastructure development. The hypothesis that construction of Konkan railway has led to higher growth of population which finds an expression in urbanization of the region is substantiated by the above analysis. During the decade 1991-2001 this region experienced average urban population growth rate (35.55%) higher than that recorded in 1981-91 (32.42%), while rural growth in 1991-2001 is lower than in 1981-91. Its urban growth is higher than the national average (31.39%) while its rural growth rate is lower than the all India average. The impact of infrastructure development is visible in the emergence of new towns on Konkan railway. Seventeen places (villages) located along the Konkan railway are designated as 'towns' for the first time in 2001 census. Towns located on Konkan railway and National Highway have recorded population growth rate higher than the region's average. The construction of Konkan railway has generated additional opportunities of development in the region which has led to migration of population, specially towards the new and existing urban centres located along the railway line. This is reflected in the sex ratio. Towns with high growth are associated with low sex ratio mainly in Raigad district. The slow growth of small towns in the interior and decline in the growth rates of rural population suggests that commissioning of Konkan railway might have helped in the outward movement of population from small towns and villages due to the increased accessibility of this region with rest of the country. To conclude, Konkan railway has accelerated the process of urbanization in the region.

REFERENCES

Asian Development Bank, (September, 1999). *Karnataka Urban Development and Coastal Environmental Management Project,* Report and Recommendations of the President.

Census of India, (1981). *Regional Divisions of India: a Cartographic Analysis,* Series 1, Vols. of Maharashtra, Goa and Karnataka.

Census of India, (2001). Paper 2 of 2001, Rural Urban Distribution of Population, Volumes of Maharashtra, Goa and Karnataka States, Office of the Registrar General of India, New Delhi.

Census of India, 1991, District Census Handbooks of Raigad, Ratnagiri, Sindhudurg (Maharashtra), Goa, Uttar Kannada and Dakshin Kannada District (Karnataka).

Environmental Master Plan Study for Dakshina Kannada District, Department of Forest, Ecology and Environment, Karnataka State and Danida, Denmark. May 1994.

Infrastructure in India, (November 1996). Economic Intelligence Service, Centre for Monitoring Indian Economy.

Krishnaveni, B. and Tapati Mukhopadhyay, (January-June 2001). Environmental degradation in newly developed urban space in Navi Mumbai—a case study, *The Deccan Geographer*, Vol. 39, No. 1, pp. 91-100.

Pilot project on natural resource accounting in Goa, (Phase 1) (2001). Final Report. Prepared for Directorate of Planning and Statistics, Goa. TERI Project Report No. 99, RD 61, New Delhi.

Shivdasani, Menka and Raju Kane, (1998). *Konkan railway, a dream come true.* Konkan Railway Corporation Ltd., Navi Mumbai.

Singh, Biri and R. Chand, (March 2000). Spatial pattern of urbanization in Himachal Pradesh, 1981-91, *Geographical Review of India,* Vol.62, No. 1, pp. 58-70.

World Development Report 1994, (1994). *Infrastructure for Development.* The World Bank, Washington D.C.

5

Tourism Development

The development of tourist industry is based on the development of the transport network. The issue of provision of efficient transport infrastructure is also closely associated with tourism development. Railways in India have been playing a vital role in tourism development. "Experience the rhythm of India—with Indian railways" is the slogan of Railways to attract tourists. The diversity of India in a true sense can be witnessed from a train. In India railways have made great strides in technology resulting in increased speed, comfort and carrying capacity. A century old "Toy Train" of Darjeeling has received recognition by UNESCO as the "World Heritage". The "Fairy Queen" steam locomotive was awarded as the most innovative and unique tourism project by Government of India. There is revival of steam heritage trains to boost heritage tourism in the country. The toy or mini trains with their serpentine tracks meandering across the hills of Kalka-Simla, Nilgiri-Ooty, Neral-Matheran are the joy rides. They are the marvels of rail engineering and spell a magical effect on tourists. Indian railways have introduced unique luxury trains— "Palace on wheels" and "the Royal Orient". The fascinating luxury saloons of erstwhile Royal houses have been converted into major tourist attraction. They combine both royal charm and modern convenience. The bullet train of Japan and TGV of France have made travel more comfortable and generated a sense of safety and environmental friendliness in the minds of the travellers. In Switzerland a trip to the top of the Jungfraujoch (at 3500 metres) the highest railway station in Europe, has long been an exciting attraction to countless travellers. Tourism is an increasingly important sector for national, regional and local economies throughout the world. In the recent times, tourism industry has been growing at a faster rate than ever before. Tourism has immense potential for boosting economic growth in developing countries. Tourists from affluent countries bring along with them the valuable foreign exchange which is very vital and also indirectly help in generating job opportunities for the locals. Geographers are interested in the spatial pattern of this expanding service sector.

This chapter highlights the role of Konkan railway in promoting tourism development and environmental conservation. It also focuses on the recent developments in foreign direct investment/collaboration projects in hotel, travel and tourism sector coming up on the western coast of India after the commissioning of Konkan railway. The study is significant because Konkan railway is a major step towards infrastructure development for promotion of tourism. It connects Mumbai, India's commercial capital and a "gateway" of tourists on the western coast and Cochin in the south through Goa, India's leading tourist destination. Opening new areas for tourism, while protecting and improving the environment poses a challenge. To what extent the Konkan Railway meets this challenge is analysed in this chapter. Our study region provides a unique opportunity to study the development of tourism in rural, mountain and coastal areas together and it is interesting to see how a newly commissioned railway line can be utilised as an effective tool for the development of tourism industry, simultaneously helping the environmental conservation of the region.

Railway and Tourism: A Historical Perspective

Introduction of railways was a crucial landmark in the world's travel history. Passengers were first carried by rail in 1830 in England between Liverpool and Manchester. The success of this railway encouraged many other countries in Europe to open the railway lines. This revolution in transport technology produced immediate expansion in European tourism. The introduction of railways vastly increased the opportunities for escape from the rigours of city life and have a boost to recreational or cultural pursuits. The railways served as the most powerful motive for mass travel in the 19th and 20th century. Rapid industrialization, urbanisation and agricultural development together increased freight and passenger transportation, and higher real income stimulated tourism related travel. Railways are likely to remain well suited through most of the 21st century specially for leisure travel. Indian railways as the lifeline of transportation have contributed immensely to the promotion of tourism.

Railways and Environment

Railway is recognised the world over as the most environmental friendly and far more energy efficient over other transport modes. Rail transportation is a boon on the background of pollution caused by the road traffic. The rail sector has the advantages over other sectors of transport in terms of cost, energy efficiency, safety and land usage. It can transport large number of people and huge quantities of goods on time, in a fast, safe and environmental friendly manner. Railways have distinct advantages in environmental protection of the sensitive regions. Environmental protection and cultural preservation is one of the thrust areas for development of tourism. Hence the Government is promoting various programmes for sustainable heritage tourism. According to the recommendation of the National Committee on Tourism 1986, tourism development needs to be harmonised with local ecological conditions. The construction of Konkan Railway was planned so as to fully preserve the ecology of the area and enhance the beauty of the environment. Extreme care was taken to avoid heritage, religious and community structures during the construction.

Tourism, Environment and Backward area Development

The increase in tourism industry has a powerful role in shaping economic development, and ecological conditions of the global environment. Tourism is being pushed as a "zero pollution industry". As the tourist industry expanded rapidly, its associated environmental degradation emerged as a major issue in several countries. The sheer number of tourists is extracting a toll on the environment in the developed and developing worlds alike as more and more land is cleared for tourist accommodation and amenities and for access roads and airports. The growing pressure of tourism and consequential development of built structures are a testimony to consumerism and lack of sustainable approach. Tourism may have destructive effects upon environmental and socio economic systems of heritage. Reducing environmental impacts of tourism has assumed highest priority in India's tourism policy.

The objective to create transport and other infrastructure for tourists coming to the region is not only to satisfy the tourists alone, but also the development of the region. Tourism contributes substantially to the economic development of backward and remote areas. Tourism development is sought as a relatively quick and simple solution to the problems of underdeveloped economies. A tourist comes to see the glories of the past and to enjoy the beauties of the present. The experience world over points out that the more prestine the natural environment, the more tourists. The more tourists, the more profit. It is a benevolent circle, a process in which both visitors and local residents ultimately win. India's tourist policy gives emphasis on developing backward areas having ancient monuments, natural scenery, festivals and local handicraft. Several tourist centres in India are located in remote and rural areas. Konkan and coastal Karnataka are such remote, rural and backward areas. Thousands of foreign and domestic tourists flock to such areas to experience the wilderness and the primeval beauty of such places. Tourism offers enormous potential for economically utilising the natural attractions of the western coast-mountains, rivers, beaches and the tranquility attached to them, and the multitude of man made attractions like mountains, caves, forts, which were constructed over the years. Konkan railways is instrumental in tapping their potential.

Konkan Railway: An Engineering Marvel and Tourist Attraction

There are numerous examples of coastline railways worldwide. Notable cases are on the Spanish, French and Italian rivieras and the coast of Britain. In Sri Lanka a railway runs on or near the shoreline for 170 km. On this background, Konkan railway an entirely coastal 760 km long railway line is unique in the world. It is the longest railway line ever constructed in one stretch in the country. Its massive railway structures are the engineering marvels and the admired features. The longest bridge on the Sharavati river, The Panval viaduct the tallest viaduct in Asia, the Karbude tunnel, the longest tunnel in Asia and in fact, the Konkan railway itself has become the tourist attraction. Therefore, those emigrated from here as well as others cherish a desire to travel by Konkan railway to see and experience these marvels. Those who have travelled testify that in Konkan railway, one just does not travel, but gets a fascinating experience. The railway line is a source of visual pleasure. For the residents of western coast the sight of a train passing is a pleasant sensation. It has enhanced the beauty of the Konkan landscape and its tourism potential. It has created among the local people a new awareness of their own heritage and environment. This awareness has also lead to the revival of traditional handicrafts with tourism development.

A New Vehicle for Tourist Movement

This entire coastal belt except Goa was untouched by railway prior to commissioning of Konkan railway. This railway line has opened new frontiers of tourism and related development for the entire western coastal region. Konkan railway is playing a key role in the integrated development of tourist destinations on the picturesque west coast of India. In a region abounding in tourist destinations Konkan railway acts as a catalyst for growth attracting both domestic and foreign tourists. A train ride through the valleys or mountains is a pleasant undertaking made easier by the absence of continuous climbing curves and gives a much smoother run. Until now only Goa had been a popular destination of foreign tourists. Konkan railway has opened new frontiers for Indian tourism as the places of tourist interest on the western coast now have a better access. The new railway line has brought destinations closer, open to a wider world. And it does it with the benefit of providing cheaper and quicker travel allowing tourists to pack more in their limited vacations. Breathtaking prestine beauty and a variety of magnificent sites greet one all along the Konkan railway route. The western coast offers tremendous variety—hill stations, historical forts—both on land and sea, sun kissed beaches, pilgrim centres, refreshing greenery. Several sights of religious and historical importance, geographical significance, geological profundity, architectural uniqueness make the journey as memorable to the tourist. Many of these spots are untouched and prestine. Until now they were inaccessible except to hardy explores willing to suffer the discomfort of exhausting bus journey. A traveller has described a journey on Konkan Railway on the high speed Janshatabdi Express from Mumbai to Goa as "one of the best journeys ever in Indian Railway in my life"(www.geocities.com/konkanrailfan,2/26/2003). Another journalist has described an exhilarating experience of his day time journey on Konkan Railway during the monsoon season as most memorable. "Every valley, every riverbank, every stretch of land unfolding after you emerge out of the tunnel reveals endless shades of green not only in the paddy fields but also on the mountains and forests" (Indian Railways, November 2000, pp.13-15).

Places of Tourist Interest Along Konkan Railway

Coastal Maharashtra

The Sahyadris are home to several charming hill stations and forts. The coastal districts are known for forts, Buddhist caves, places of pilgrimage and the beaches. Raigad fort is the crown of all of Shivaji's forts. It is 35 kms from Veer station of Konkan railway and is accessible by ropeway. Shivaji's coronation took place here in 1674. Market place, Mahadwar, Gangasagar lake, Raj Sabha are the places of tourist interest. Alibag, is a convenient base for visiting Kihim and Nagaon beaches and the Kolaba sea fort built by Kanhoji Angre. Kihim is an exclusive beach laced with lush coconut trees. Nagaon and Akshi beaches, secluded and lined with casurina forest are enchanting places for a day's outing from Alibag. There are temples of the time of Peshwas, a Jewish synagogue and an observatory. Formerly the capital town of the Siddis of Janjira, Murud is popular for its alluring beach and a 300 year old Janjira sea fort. The fort enjoyed an excellent command over the Arabean sea. Harihareshwar, known for its tranquil and picturesque beach, is also known for its Shiva temple and is called as Dakshin Kashi. The nearest railway station is Mangaon (65 km) on Konkan Railway Srivardhan beach with soft sands and inviting waters is 19 km from here. Kuda caves is a group of 26 Buddhist caves and 11 cisterns. These caves command a beautiful view. There are some inscriptions in the caves still legible. Pale is a group of 29 Buddhist caves. Samartha Ramdas stayed in Shivthar cave and wrote *Dasbodh*, a literary composition. His Samadhi and full sized idol are located here (Fig. 5.1).

Sav hot springs (a group of three springs) are on the bank of river Savitri. Water is insipid and sulphurous. People suffering from skin diseases take bath in the springs. Pali also has hot springs possessing medicinal value. Ratnagiri on Konkan railway is a charming seaside town with a backdrop of a 15th century fort. It is famous for Alfonso mangoes, cocoanuts, jackfruit and a long stretch of black sandy beach. This is the birth place of Lokmanya Tilak and his memorial is maintained. Chiplun on Konkan Railway has emerged as a major industrial centre. Ganpatipule 19 km north of Ratnagiri is the 400 year old temple of Lord Ganesh. It is built in a colossal rock on the shore and the idol is believed to be "Swayambhu." There is a beautiful unspoiled beach of white sand with a picturesque and quiet surrounding. The place is developed as a tourist resort. Nestling in the ranges of Sahyadri at an altitude of 690 metres, engulfed by dense forest. Amboli is a splendidly beautiful mountain resort in Sindhudurg district. With its salubrious climate, spectacular waterfall and the mist clad Konkan at the bottom of the hills, Amboli presents a panoramic views of the region. It is the most visited hill station in Western Ghats. Sawantwadi on Konkan railway is a convenient railhead to reach this place (28 km away). Sindhudurg and Vijaydurg are the famous sea forts built by Shivaji. Vengurla has a beautiful sea shore. A coastal stretch of 84 km is ideal for beach tourism. It is declared as "special tourist area". Sindhudurg district is identified as an important "tourist destination" for development through joint efforts of Central and State Govts. Raigad and Janjira fort, Kuda caves, Harihareshwar and Sindhudurg form a travel circuit identified for development under the Action plan based on "Travel circuit concept" by Ministry of Tourism, Government of India. Tata Consultancy Services have identified 22 places in this district for development. They include 6 beaches, 5 religious places, 4 heritage and cultural sites, two lakes and a hill station. Five travel circuits within the district are identified. Construction of an airport is also proposed.

Tourist Attractions in Goa

Goa a major beach resort is a popular hotspot where domestic and foreign tourists make a bee-line throughout the year. The rail distance between Mumbai and Goa is 190 km less than the road distance. Goa has a distinct culture and tradition because of its former Portugese rule. The world famous beaches of Gao present the most unforgettable sights. Madgaon, known as Gateway to Goa, is the cultural centre and the most prominent junction on Konkan railway. Just 6 km from Madgaon station of Konkan railway is the splendid 18 km stretch of silvery sand—the Kolva beach. The Calangute beach north of Mandovi river, Anjuna, Miramar, Vagator, Benaulim, Dona Paola and Palolen beaches offer a breathtaking panorama. Prominent temples in Goa are Shantadurga, Mangeshi, Mahalasa, Bhagavati, Ramnath, Mallikarjun, Brahma and Tapobhoomi. Imposing lamp towers in front of every temple is a unique feature in Goa. Goa has thick forests and a variety of wild life. There is a jungle resort at Bondla forest. The wild life sanctuaries are at Bondla, Cotigao and Molem and a Bird sanctuary at Chorao island. Mayem lake is a beautiful picnic spot.

Panaji, the capital of Goa is located at half an hour's drive from Karmali station of Konkan Railway.

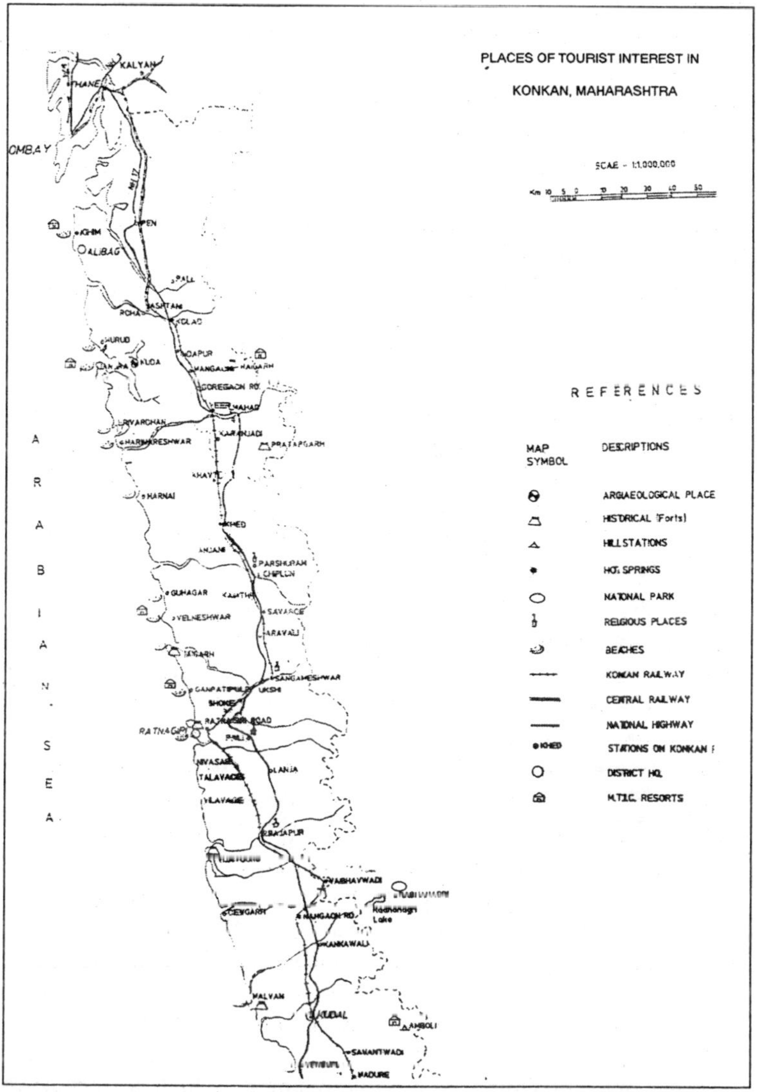

Fig. 5.1 Places of Tourist Interest in Konkan, Maharashtra

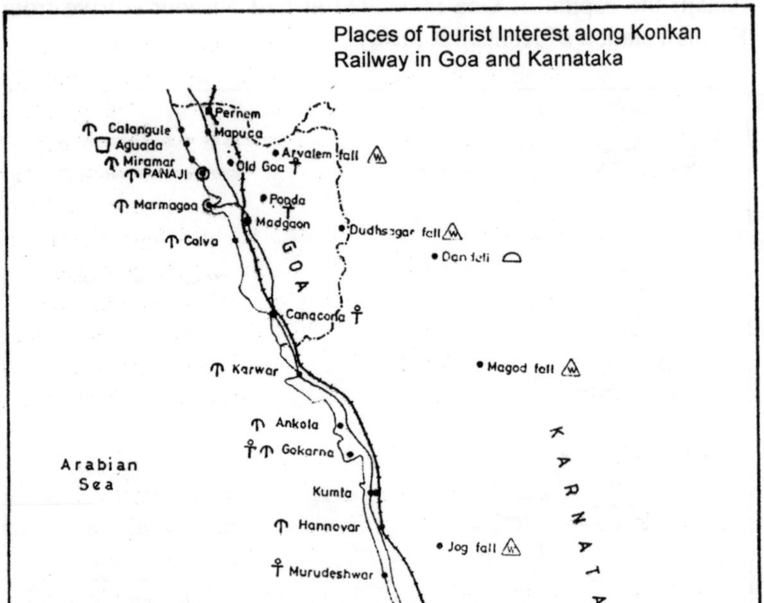

Fig. 5.2: Places of Tourist Interest along Konkan Railway in Goa and Karnataka

It has majestic churches, prominent among which is Bom Jesus Basilica, built in 16th century, having the embalmed body of St. Francis Xavier. Old Goa can also be called as the city of gigantic churches. Its cathedral has a large bell. The church of St. Francis of

Assisi houses a museum. Goa is dotted with historic forts. The famous ones are Aguada, Terekol, Chapora and Cabo de Rama. Mormugao is India's major port on the mouth of river Zuari, known for its export of iron ore. It is connected to Konkan Railway at Madgaon station by a branch line of South Central Railway. The Tourism Department of Goa carries out illuminations on sites of tourists interest like Raj Bhavan, temples and churches. Tourism is the major economic activity in coastal tahsils of Goa. The statistics of tourists arrivals in each tahsil over the years and their increasing trend point out the land conversion (land put to residential and non-residential use). Increase in number of hotels, guest houses and other facilities for tourists is also evident from Goa's Accommodation Directory published by the Directorate of Tourism. It provides an indirect evidence of changes in land use in Goa. Konkan Railway has helped in boosting tourist traffic in Goa (Fig. 5.2).

Coastal Karnataka

Karwar, once an important trade centre, has a famous Devbagh beach, protected port and a naval base is coming up. 60 kms from Karwar is Yana famous for its unique and gigantic rock formation of a height of 90 metres. Encircled by thick forests, it is a favourite haunt for rock climbers. A few kilometres away from the Dandeli Sanctuary, is a beautiful cave perched in the middle of a steep rock, with a large Shivalinga—like formation formed by the stalactites and stalagmites. Gokarna town is formed by the 'ear-shaped' confluence of 2 rivers. It is believed that the place derives its name due to Shiva emerging from the ears of a cow. Gokarna is famous for the 'Atmalinga' in the ancient temple here. Gokarna is a famous Sanskrit learning centre. The Om Beach and the Kutle Beach nearby are also worth a visit. Murudeshwar a picturesque place situated 16 km from Bhatkal is bounded by the shimmering sea and rolling hills. It has a gorgeous temple perched on a hillock by the shore. The temple which embodies a linga is believed to have erupted when Ravana flung the cloth covering the atmalinga at Gokarna while lifting it. On the hillock there is a shrine of Jattiga and a huge fort, said to have been renovated by Tipu. Jog (60 km from Honnavar), is famous for its magnificent waterfalls. The Sharavati river flowing over a rocky bed takes a spectacular leap from a height of 292 metres and divides into 4 smaller ones—known as the Raja, Rani, Rocket and Roarer—presenting a glorious view. A hydroelectric station nearby supplies electricity to Karnataka. About 80 km fom Udipi in South Kannada district is the captivating town of Agumbe situated atop a ghat head. It receives the highest rainfall in Karnataka and has many scenic delights to offer. The glorious sunset is the main attraction of this place.

Agumbe, a captivating town situated atop a ghat head, receives the highest rainfall in Karnataka. This place has many scenic spots. The glorious sunset is its main attraction. The Gopalkrishna temple and the Observatory are the other places of interest. Situated at the foot of a high rock, forming a part of the Kudremukh range of hills, Jamalabad is famous for its Fort built by Tipu Sultan in 1794. About 80 km from Udupi the Mookambika Temple at the foothills at Kollur attracts thousands of devotees. The idol that was installed by Adi Shankara is confined in a copper-roofed and gold crested temple. Arasina Makki is a famous waterfall nearby. Udupi, 60 kms. from Mangalore, was the 'sanctum' of Madhavacharya, the great Sanskrit philosopher. The famous temple here has an idol of Lord Krishna richly adorned with jewels. The main attraction of this temple is the 'Kanakana kindi'—a small window through which Krishna is believed to have given *darshan* to his ardent devotee, Kanakadasa. Udupi is known for its mouth watering cuisines. Malpe, 6 km from Udupi, is an ideal picnic spot with its virgin beach. It has the endless stretch of golden sand, graciously swaying palm trees, the clear blue sky.

St. Mary's island is a delightful getaway spot right across Malpe. Vasco-de-Gama landed here in 1498. About 27 km east of Mangalore at Katil, on the banks of the river Nandini, there is a famous temple dedicated to Goddess Durga Parameshwari. Karkala has the massive 12.8 m tall monolithic Gomateshwara statue and the St. Lawrence Church. The town is known for its sculptors. Statues made here are widely exported and appreciated the world over. Dharmshala has a 14 m monolith of Bahubali, and the Manjunatha temple. The Subramanya Temple was once the refuge of the Nagas, the cave dwellers. The temple has an idol of Lord Kartikeya, worshipped in the form of a cobra. Mangalore is a city of vivid contrasts. It has narrow winding streets, lofty coconut trees, beautiful beaches, temples and churches, the heady fragrance of the Mangalore Mallige, the aroma of spicy coconut curries. Mangalore is a major port and an industrial centre with oil refinery, chemical fertilizers and iron and steel industries.

National Parks and Wild Life Sanctuaries

- *Maharashtra*: Karnala bird sanctuary is 12 km away from Panvel railway station north of Roha on Konkan railway. The sanctuary provides shelter to a variety of wild life. A fort and trekking facilities have made it popular among tourists. Dajipur and Radhanagari Bison Sanctuary located in adjoining Kolhapur district is accessible by Sawantwadi railway station of Konkan railway. Apart from Bison, it is also rich in other wild life. Forest parks maintained by the State Forest Department in coastal districts are at Amboli, and Narendra dongar in Sindhudurg district and Matheran in Raigad district.
- *Goa*: There is one national park and six wild life sanctuaries in Goa. Chorao island at the mouth of river Mandovi with a criss-cross network of water channels and tidal variations has naturally regenerated mangrove vegetation with a variety of mangrove species. A mangrove nursery was established here. Of this, an area of 1.8 sq. km was declared as bird sanctuary and named after Dr. Salim Ali, the renouned ornithologist. During winter thousands of migratory birds flock here and feed on the banks of river adjoining the sanctuary.

  Bondla is a small sanctuary just 8 sq. km located among the hills near Ponda. Bondla is a paradise for eco tourists. It has a mini zoo, deer safari park, botanical garden, nature education centre, nature trails and eco tourism cottages. The area is covered with moist deciduous forests with small patches of evergreens and canes. The thorny bamboo abound. Located at the foothills of Western Ghats, Bhagwan Mahaveer wild life sanctuary and National Park at Molem is the largest sanctuary of Goa. This encompasses 133 sq. km of rich forests varying from moist deciduous, semi-evergreen and evergreen. Nature Information Centre at Molem, a 13th century Mahadeo temple 13 km away, Dudhsagar falls, the highest waterfalls of Goa and Devil's canyon nearby are the other places of tourist interest.

Located near the Canacona railway station of Konkan railway (9 km) in South Goa, Cotigao Wild Life sanctuary is the oldest protected area of the state. It spreads over an area of 85.65 sq. km intersected by Talpona and Galgibag rivers. Fully covered with dense forests, it is representative of the ecosystem in the Western Ghats region. Large portions have a forest crown density of over 50 per cent. The vegetation is of moist deciduous type interspersed with semi-evergreen and evergreen patches, cane and bamboo. Inside the sanctuary are water holes, artificial salt licks near the water holes and watch towers. All these water bodies attract the animals during the summer months. Large number of animal species and about 200 bird species could be seen. Netravali wild life sanctuary connects to the Bhagwan Mahaveer wild life sanctuary on the north-west border. This is a forest corridor along which animals move freely.

- *Karnataka*: In Uttar Kannada district 157 km away from Karwar station lies Dandeli—a natural habitat for wild life set against the backdrop of Western Ghats. The famous Dandeli game sanctuary houses some rare species of wild animals like barking deer, sloth bear, Sambar, Gaur and elephant. Jog waterfalls has plenty of wild life.

The future of tourism along the western coast depends on the prospective for promoting wild life eco tourism. It was widely apprehended that wild life, reptiles and birds may get affected due to disturbances as a result of human and machinery movement during construction and noise pollution during operation. However, the Konkan railway line does not pass through any sanctuaries, but passes at a distance of 4.5 km away from south-western border of Cotigao wild life sanctuary and hence does not pose any threat to wild life.

Beach and Cruise Tourism

Beach tourism in this region is expected to get a huge boost with the Union Tourism Ministry asking the World Tourism Organisation to conduct a feasibility study on utilizing the west coast for cruise tourism. The five ports identified for this study are Mumbai, Goa,

Mangalore, Cochin and Tuticorin. The infrastructure and carrying capacity of these five ports will be studied. Cruise tourism is becoming popular and recently the number of tourists undertaking cruise tourism has increased. It has become a huge revenue generator and the main income source for islands such as the Caribbeans. However, cruise tourism could not develop in India due to lack of facilities. The WTO study will focus on whether the ports are tourist friendly and suggest what requires to be done in this direction. The Tourism Ministry selected the west coast due to its proximity to Europe and Africa, where many of the cruise tourist companies operate.

Package Tours

Indian Railways in collaboration with the State Tourism Development Corporation have introduced short package tours called as "Rail Weekenders." These weekend packages catering to special interests take to various historical places, cultural and pilgrim centres. To enhance tourism KR has tied up with reputed tour operators and selected resorts for a complete tour package. Certain hotels in Goa are also identified for these tour packages. These package tours started since October 2000. Total number of packages booked by the resorts during the year 2000-01 was 520. Mumbai Madgaon Mumbai confirmed booking to tourists is made available by KR in 3 tier AC coach. Package tours are also available between Mumbai and Ratnagiri by Mandovi Express and Jan-Shatabdi Express. The weekend tours from Mumbai to Goa offer a welcome respite for the urban dwellers. New traffic is attracted by package tours. Maharashtra Tourism Development Corporation (MTDC) had plans to deploy "Hotel on wheel" concept on Konkan railway. The Railway Minister in the Budget speech on 25-2-2000 announced an introduction of a new train Deccan Odyssey, a luxury train on the lines of Palace-on-Wheels of Rajasthan to connect tourist places in western and coastal Maharashtra. This joint venture of Indian Railways and MTDC has materialized and the Deccan Odyssey completed its maiden seven day journey on 5th February 2004. This fully air conditioned 21 coach sleek blue train or a five-star hotel on rails with an investment of Rs. 32 crore is on par with the blue train of South Africa and the Oriental Express of Europe. This 150 kmph speed bullet train is equipped with gymnasium, a beauty parlour, an ayurvedic massage centre, restaurant, bar and a conference room. It is being promoted at a travel fair mart at London and Germany. The train offers tourists a glimpse of the cultural beauty of Maharashtra. Starting from Mumbai it covers the heritage and the tourist sites such as Ajanta, Ellora, Maharashtra's cultural capital Pune, famous wineries of Nasik. The train has two stops along the Konkan coast allowing tourists to discover the backwaters of Maharashtra and the golden beaches of Ganpatipule and Tarkarli beach resort and visit the spectacular Sindhudurg fort. With this, the whole of Konkan region is opened up to foreign tourists. MTDC is planning to offer short packages of three days during the off-season period to Indian tourists. Sindhudurg district is declared as a "Tourist District" by Government of Maharashtra and Maharashtra Tourism Development Corporation is planning to introduce special houseboats depicting the Konkani culture at selected beaches in Ratnagiri and Sindhudurg districts. Foreign tourists coming by Deccan Odyssey would be welcomed at Sangameshwar railway station and taken to local sight seeing by an air conditioned bus and can stay and move along the coast by these house boats. This project is expected to get a good response not only from the foreign tourists, but also from tourists of North India who have a tremendous attraction of the sea and beaches. Major hotel chains in Goa are expected to operate a four coach tourist specials during day time from Vasco or Madgaon to tourist spots like Dudh Sagar water fall, Canacona, Balli and other spots. With this, the immense tourist potential of Goa could be exploited further. Foreign tourists in Goa amount to 11.6 per cent of total foreign tourists coming to India and charter tourists to Goa account for 75 per cent of total charter traffic to India. Goa received 2.4 lakh foreign tourists in 1996 which constituted 10 per cent of the total number of foreign tourists that arrived in India. The contribution of the tourist industry to the Goan economy has increased substantially. In 1994 the contribution was 11 per cent of the state GDP. According to the survey conducted by NCAER, total estimated domestic tourists in 2002 at the national level are about 550 million, of which about 29 per cent were urban and 71 per cent rural. Maharashtra ranked first and Karnataka third in India according to their relative share in total tourists. Total expenditure incurred by domestic tourists in the country during 2002 is estimated to the tune of Rs. 319 billion (PIB Press Release, Ministry of Tourism and Culture, 12 March, 2004) These figures reflect the growing importance of domestic tourism and the contribution of Konkan Railway towards the same.

Passenger Amenities

Internet booking is a major step taken by Konkan Railway to cater to foreign tourists. This is for the first time that such a facility has been started in India, and a landmark in providing innovative packages for the convenience of passengers. It is available for 3-tier AC berths in Konkan Kanya Express between Mumbai and Madgaon (Goa). The ticket is handed over to the passengers at the train itself on producing copy of e-mail confirmation sent by KR (KR Bulletin Board, 28-2-2002). The booking is available from 7 days before the date of journey. This shows the growing popularity of Konkan railway among foreign tourists. At Madgaon railway station there is a separate ticket booking counter for foreign tourists. Considering the demands of passengers, "Tatkal service" (an instant confirmed reservation on demand) for sleeper class is available for Konkan Kanya Express. In Konkan railway the entire information dissemination for passengers is controlled by the computer. Hence passengers are able to obtain services much more satisfactorily from a Konkan Railway station. Konkan Railway launched the first cyber café centre at Ratnagiri station on February 1, 2004 to make information available to the public and be a catalyst in the all round growth of the region. The cyber café offers facilities like audio video chats and printouts. Here the passengers would be able to surf websites and send e-mails, keep track of the train, check ticket status and reservation availability through Konkan Railway website. Two more cyber cafés are in the pipeline and will be commissioned in the near future. Konkan Railway has taken adequate care to ensure that even small stations are provided with the minimum basic amenities like drinking water, tea stall, waiting hall, retiring room, parcel booking office. Restaurants serve delicacies to passengers on the platform. Mobile trolleys serving snacks, fruits and curio are provided at various stations in addition to stalls. Additional amenities like water coolers, foot over bridge (FOB), wheel chairs, stretchers for invalids and the sick have also been provided at major stations. Drinking water amenities are available at all stations. Better maintenance of railway stations, better house keeping services in AC coaches, are the other facilities the Corporation has introduced to promote tourism. Cleaning of stations is undertaken at all the stations regularly. "Railotel" budget rail hotel has been commissioned at Kumta near Karwar. Some more "Railotels" are also coming up. For the convenience of tourists, the timetable of Konkan Railway includes a list of places of tourist interest on the western coast. The distance from the nearest station of Konkan railway is also mentioned. Tourist information station-wise and information about its passenger train punctuality is available on internet on website of Konkan Railway. The Tourism departments of all the three coastal states through which the Konkan railway passes, i.e. Maharashtra, Goa and Karnataka have their websites to provide the detailed tourist information.

Integrated Road Rail Train: Konkan Railway proposes to operate integrated road rail train services between metropolitan cities to generate traffic by competing with roadways. An integrated road rail service provides door to door service and can open entirely new market of high rated traffic, and this will be an added value service to the customers (Directors' Report, 10th Annual Report, KRCL,1999-2000, p. 7).

Phases of Tourism Development

The evolution of tourism or stages of tourism and infrastructure development in western India are comparable with similar studies undertaken in other parts of the world. In a study of evolution of tourism in Scottish Highlands, Butler (1986) has identified distinct phases of tourism development. In the earlier phase travel is difficult and there is lack of adequate accommodation and transportation. Gradually there is a growth in visitor numbers, travel becomes easier and guidebooks on the region begin to be published. This process is accelerated in the age of "railways and hotels". The expansion of the railway system is followed by a wave of hotel construction and the emergence of resorts (touring centres). Travel now becomes highly accessible for the "active tourists" or middle classes. Finally, in the age of automobiles, the region becomes a popular playground for car owners. This brings about intense environmental pressures and cultural challenge as the number of tourists and in-migrant families increases. The western coastal region also seems to be undergoing through similar phases of development.

Investment in Hotel Industry after the Completion of Konkan Railway

The location of the towns in relation to Konkan railway and National Highway as shown in the maps (Figs. 4.10 to 4.13) provide insight into growth of tourism and the way the hotels and tourist resorts are dotted along the National Highway. Konkan Railway in the long term perspective has become complementary. On the western coast of India only Goa had remained a popular destination for both domestic and foreign tourists since long. Due to lack of transport infrastructure there was no significant development in the "approved" hotels category (approved by Ministry of Tourism, Government of India) or other tourism infrastructure in coastal Maharashtra or Karnataka. The roads leading to tourist destination did not have adequate facilities like hotels and restaurants. After the beginning of Konkan Railway construction there was a wave of construction activities in this region. New hotels are under construction at several places along the Konkan Railway route. Maharashtra Tourism Development Corporation has constructed sea side cottages and tent resorts at Ganpatipule, Tarkarli, Harihareshwar, Raigad. Central Ministry of Tourism is also providing financial assistance to State Government for this purpose. Such financial assistance was not received in this region prior to construction of Konkan Railway and has rushed in only after its commissioning. Mangaon is a small station on Konkan Railway in Raigad district. and the establishment of hotel project by Suman Hotels at Mangaon is directly attributed to Konkan railway. Konkan Railway invited tenders in December 2000 for provision of railotels at 16 stations of KR to improve passenger amenities. KRCL will offer plots in the vicinity of stations for providing Railotels on lease rent to construct and operate a hotel with minimum 20 rooms at 16 stations, viz. Kolad, Veer, Chiplun, Ratnagiri, Kudal, Sawantwadi, Thivim, Karmali, Madgaon, Canacona, Karwar, Murdeshwar, Bhatlkal, Kundapura, Udupi and Surathkal.(Konkan Railway Press Release) Karnataka government is trying to attract private investment to set up hotels and resorts on select places identified on its 300 km coastline and in its lush forest zones. Jungle Lodge and Resorts is promoting eco tourism in the state and planning to open eco tourism resorts at Tannir Bhavi and Belikeri near Ankola (a station on Konkan railway) both on the coast and near Jog falls to construct house boats. Eco tourism is also promoted at Dandeli, accessible from Konkan Railway. Due to infrastructure bottlenecks, i.e. with link roads in bad shape, no big promoters want to put money into hotels in this region (*Times of India*, 16-3-2001). Under the new joint venture company for expediting railway infrastructure projects in Karnataka, the coastal towns of Ankola and Mangalore would be connected to the interior towns of Hubli and Hassan by new broad gauge railway line.

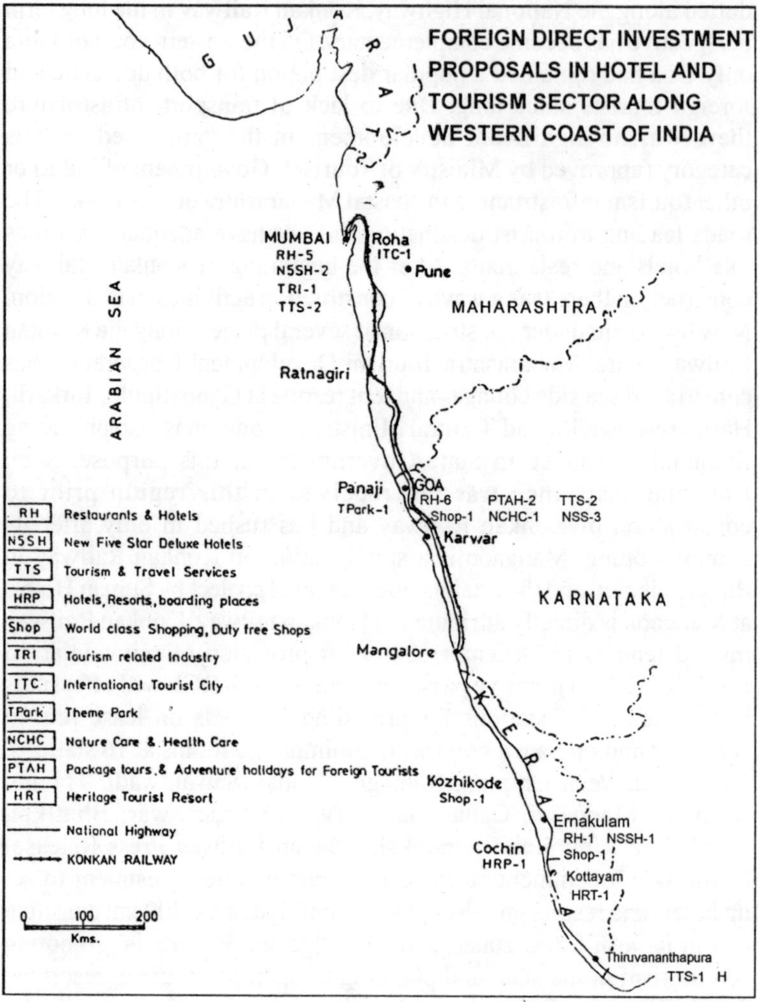

Fig. 5.3: Foreign Direct Investment Proposal in Hotel and Tourism Sector along with Western Coast of India

Foreign Direct Investment in Hotel and Tourism

After the commissioning of Konkan Railway, this region has emerged as the focus of attraction of foreign direct investment in travel and tourism sector in India. Since the commissioning of Konkan Railway in January 1998 to April 2000, out of 77 proposals approved in this sector in all India, 36 are located on the western coast (Fig. 5.3). This clearly shows the impact of Konkan Railway. Konkan Railway is indeed instrumental in promoting tourism and attracting foreign direct investment in tourism sector. Among the states which are the beneficiaries of Konkan Railway, Goa has attracted the largest number of proposals (15) followed by coastal Maharashtra (12) while coastal Karnataka has failed to attract any proposal. Apart from the total number of proposals it is equally important to evaluate the share of hotel and tourism sector in total FDI projects in each state. This sector accounts for one-third of total FDI proposals in Maharashtra and Goa. Goa attracted a total of 45 proposals in various sectors from 16 countries. Among them 15 are relating to hotel and tourism. They include establishment and operation of two Five Star deluxe beach resorts, hotels, rooming houses, camps, lodging, operating modernized restaurants, food stores, coffee shops at hotels and resorts, package tours and adventure holidays for foreign tourists, to provide world class shopping and logistic services, nature care and health care and theme park. Collaborating countries are USA, UK, Germany, Mauritius, France, Switzerland, Brazil and Panama. On coastal Maharashtra the bulk of FDI proposals are concentrated in Greater Mumbai. They include two new Five Star deluxe hotels, five restaurants and hotels, two tourism and travel related services and one travel related industry. A mega project for establishing an international tourist city is proposed in Raigad district with an investment of Rs. 4125 million. Three projects completed in Goa include The Taj Exotica (October 1999), Marriot International and Mahindra Holidays and Resorts (in December 1999).

Konkan Railway: An Asset for Tourism Development

Tourism is going to stay with us. People are on the move in much greater number than ever before. Its volume shall increase as the pace of industries goes faster. Increase in the volume of passenger traffic on rails and roadways is indicative of the size of domestic tourism in the country. For the people of Konkan and the coastal region, it is not the statistics that matters. For them Konkan Railway means a new beginning, more tourists, more jobs, more income and rapid economic growth. In Goa due to increasing demands on resources and infrastructure coupled with pollution and destruction of the natural environment, the environmentalists are demanding to put limits to the expansion of tourism sector in Goa (Goa Foundation Report).

For further promotion of tourism, there is an urgent need to upgrade other infrastructure facilities for tourists, i.e. all the beaches and other places of interest in the region should be well connected by all weather metalled roads to the nearest stations of Konkan Railway and excellent public transport facilities should be made available at the railway station to passengers alighting by Konkan Railway to visit the tourist place without wasting much of time. This is particularly important in Konkan region as the railway stations are slightly away from the main cities and towns and also away from the beaches and forts compared to Goa and Karnataka. Unless a proper connectivity between them is established, Konkan Railway cannot serve its purpose of tourism promotion. Konkan Railway has to be alert to meet the challenges posed by automobile industry in the future. Konkan Railway has completed over five years of its full length operations and has made an indelible impact on tourism. This welcome addition to western India's infrastructure has proved to be an important asset for tourism development and environmental improvement.

REFERENCES

Aitken, Bill, *Exploring Indian Railways*, Oxford University Press. Delhi 1994

Annual Report, 1998-99, Ministry of Tourism, Government of India, New Delhi.

Bhatia, A.K. *Tourism Development—Principles and Practice*, Sterling Publishers, New Delhi, 1982.

Butler, R.W. Evolution of Toursm in the Scottish Highlands, *Annals of Tourism Research,* Vol. 12,1985, pp. 371-91.

Carpenter, T.G. *The Environmental Impact of Railways*, Wiley, Chichester (England) 1994.

Discover Karnataka's Coastal Circuit, Brochure Issued by Karnataka Tourism,

Ellis, Roystone, *India by Rail*, Brand Publication, London, 1989.

Gazetteer of India, Maharashtra State, *Kolaba District Gazetteer*, Revised Edition, Mumbai: 1984.

Government of Karnataka, Bangalore.

India by Rails, Department of Tourism Government of India, 1996.

India Tourism Statistics, 1998 Ministry of Tourism, Government of India, New Delhi.

Konkan Railway's website www.Konkanrailway. com.

Monthly News letters of Indian Investment Centre, New Delhi, Monthly issues from 25 April 1998 to 25 April 2000.

Shaw, Gareth and A.M. Wiliams, *Critical Issues in Tourism, a Geographical Perspective*, Blackwell. Oxford. UK, 1994.

Shivdasani, Menka and Raju Kane, *Konkan Railways: A Dream Comes True*, Konkan Railway Corporation Ltd., Navi Mumbai, 1998.

Singh, Tejveer, *et at*. (ed) *Tourism Environment: Nature, Culture, Economy*, Inter India Publication, New Delhi, 1992.

Tourism Policy of Government of India. Lok Sabha Secretariat, New Delhi, 1985.

Tourist Maharashtra: A Guide to Places of Interest in Maharashtra, Maharashtra Tourism Development Corporation, Mumbai, 1984.

6

Investment in Infrastructure and Industry

Infrastructure services like transport and ports work as intermediate inputs to industry. Improvements in these inputs in any geographical location attract flow of additional investment (private/public investment) from both domestic and international sources. This also raises the productivity of other factors of production (labour and capital) and profitability of the other producing units thereby permitting higher level of output, income, and/or employment. New industries tend to locate where land is available and infrastructure exists. Any location not served by a railway line is at a disadvantage and is not likely to attract any investment for its development. Railways have facilitated the development of industries and assisted the economic progress of the country towards self-sufficiency. The Konkan railway has opened up a vast backward area for industrial development. It has enhanced the infrastructure facilities, which in turn have encouraged the investors to venture into new industries. The Konkan railway is expected to promote further urban growth and industrial expansion. How far the coastal region has succeeded in attracting investment, for setting up of industries from private and public sector and Foreign Direct Investment (FDI) after the construction of Konkan Railway is examined in this chapter. This chapter analyses the distribution of industries on the Western coastal region prior to commissioning of Konkan railway. The industrial development projects implemented in this region after the commissioning of Konkan railway have also been considered. The analysis is based on the data collected from the office of Maharashtra State Pollution Control Board, Mumbai and National inventory of large and medium industries compiled by Central Pollution Control Board (CPCB), New Delhi.

Industrial Situation Prior to Construction of Konkan Railway

The Western coastal region is endowed with a variety of mineral and forest resources. These provide raw material to industries based on these resources. There is lot of untapped potential in the region waiting for exploitation. Lack of infrastructure facilities mainly in the area of transport prevented large parts of the region from coming into their own. As there were no rail facilities in the coastal belt, it hindered the growth of industry in the region. A variety of small and medium scale industries developed here had to depend on roads to get the raw materials and to market their products. As rail transport is now available, industrial growth based on local raw material is expected to increase and this will boost the employment potential. Faster industrial and commercial growth is expected to take place in future in the hinterlands of port towns.

Large and Medium Industries

Central Pollution Control Board carried out a countrywide survey in 1995 to ascertain the status of pollution control measures in 23 categories of large and medium sized industries causing air and water pollution.

The industries are classified on the basis of their capital investment, those having investment over Rs. two crores are classed as large and between Rs. 65 lakhs and two crores are classed as medium. This survey has been a national inventory and includes a vivid account of industrial situation of the Western coastal region prior to commissioning of Konkan railway. As per this inventory, there are 287 polluting industries in our study region, 92 among them (about one-third) are large and the rest are medium. The distribution of these industries in the region as given in the Survey is shown in Table 6.1.

Table 6.1: Large and Medium Industries on Western Coast

Type of industry	*Raigad*	*Ratnagiri**	*Goa*	*Uttar Kannada*	*Dakshina Kannada*
Chemical	51	18	2	2	—
Petrochemical	6	1	2	1	3
Fertilizer	5	—	1	—	1

(Contd.)

(Contd.)

Type of industry	Raigad	Ratnagiri*	Goa	Uttar Kannada	Dakshina Kannada
Dyes	17	2	—	—	—
Pharmaceutical	12	3	2	—	1
General Engineering	8	4	—	4	9
Rubber products	2	1	1	—	5
Food stuffs	15	1	8	3	18
Pulp and paper	7	1	—	1	—
Iron steel, minerals	11	2	2	2	4
Textile	11	2	—	—	—
Miscellaneous	6	2	3	9	15
Total	151	37	21	22	56
Large/Medium	57/94	10/27	9/12	5/17	11/45

* As only one large industry and no medium industry is enumerated in Sindhudurg district, this district is not separately shown in this table, but clubbed in Ratnagiri district.

Source: National inventory of large and medium industries and status of effluent treatment and emission control system, Vol. I&II, Central Pollution Control Board, Ministry of Environment and Forests, Government of India, New Delhi, 1997.

Industries in Coastal Maharashtra

The Konkan region remained industrially backward due to lack of infrastructure facilities, viz. absence of railway line, inadequate air connectivity, lack of trained manpower. Provision of infrastructure for industrial development is the responsibility of the government. The State Government recognized Ratnagiri and Sindhudurg as industrially backward and announced several incentives that the districts could attract new industries. Because of its location adjacent to Thane, Pune and Greater Bombay and availability of infrastructure facilities, Raigad district is marching ahead in industrial development. Mumbai Thane industrial belt reached saturation point and needed dispersal of industries. Both Raigad and Ratnagiri districts are acting as receptacles of the spill over. Raigad district is more favourably placed with its relatively less undulating territory, more fertile soil, and proximity to the offshore points-gas reserves. Ratnagiri's misfortune lies in its endless chain of mountains and innumerable rivers and streams which make establishment of communication links difficult. Its greatest asset is the hardy nature of its people.

Table 6.2: Major Industrial Companies Located on Western Coast

Ownership/group	*Industry*	*Location (District)*	*Est. sales Rs. crores*
Reliance Industries	PSF, PFY, PTA	Raigad	3436
RCF (Public sector)	Fertilizer	Raigad	1045
Hindustan Lever	Soaps & detergents	Raigad	961
Indian Petro-chemicals (Public Sector)	Petrochemicals	Raigad	931
Castrol India	Lubricants	Raigad	691
Raymond Singhania	General Engineering	Ratnagiri	918
Kudremukh iron ore (public sector)	Iron ore concentrate	Dakshin Kannada	414
Mangalore Chemicals & Fertilizers.	Fertilisers	Dakshin Kannada	220
West Coast Paper Mill	Paper	Uttar Kannada	264

Source: Profiles of States, CMIE, Mumbai, March 1997.

Raigad District

This district has the largest number of large, medium and small industrial units in the western coastal region. Over half (53%) of the region's total industries are concentrated in this district. 62 per cent of our study region's large industries and 48 per cent of medium industries are located in this district alone. 38 per cent of total industries in Raigad district are classed as large. One-third of total industries in Raigad district belong to the category of chemical industries, another one-third consists of dyes, drugs and pharmaceuticals, petrochemicals, fertilizers. The largest investment of Rs. 14,887 lakhs has been made in Chemical industry, followed by Rs. 14,749 lakhs in Engineering industries. The variety of industries found in Raigad district is simply tremendous ranging from textiles, paper and pulp, general engineering, iron and steel, foodstuffs, breweries, electronics, electrical, glass, rubber, power generation, and lubricant oil. Construction of two giant fertilizer plants at Thal Vaishet was the most important development in the district. In addition, establishment of Nhava Sheva port (now renamed as JNPT) new township with an industrial area in the neighbourhood also opened up immense possibilities of growth of traffic. There are 9 large scale public sector enterprises in Raigad district with a capital investment of Rs. 2500 crores. They are located at Rasayani, Thal Vaishet and Nagothane and provide large scale industrial employment. Indian Petrochemical Corporation at Nagothane utilizes crude oil and natural gas of Bombay High in its gas cracker plant to obtain various products.

This project has the capacity to employ about 1800 to 2000 skilled and unskilled workers. Rashtriya Chemicals and Fertilisers plant is located at Thal Vaishet with a capital investment of Rs. 1200 crores. Suphala and Ujjwala brands of Ammonia and Urea fertilizers are manufactured here. This plant is expanded with the establishment of two more units. Nippon Denro Company's sponge iron plant with a capital investment of Rs. 800 crores was commissioned at Dolvi (Tahsil Pen) in 1994-95. Taking into consideration the increasing requirement of power for industrial development, a gas based power project in private sector by Reliance Industries Ltd. is approved to produce 410 MW of electricity at Patalganga (Tahsil Panvel) with a capital investment of Rs 1380 crores. In addition to above, 281 large and medium scale projects with an investment of Rs. 21,576 crores have been cleared and under implementation. A total of 1757 small-scale industrial units are registered in the district. The industries are scattered in small villages and towns in all Tahsils of the district. However, MIDC's industrial estates at Dhatav (22), Mahad (21), Taloja (20), Patalganga (19), Roha (15-number of industrial units) represent the largest concentration of industries, about two-third of the district's total units. Khalapur, Khopoli, Panvel, Uran have emerged as major industrial areas. According to Annual Survey of industries 1998-99, 32 per cent of industries in the district (268) employed less than 10 workers each, 66 per cent (552) employed 10 to 499 workers, and only 1.8 per cent (15) industries employed over 500 workers. The impact of industries on population growth, urbanization and emergence of "new" towns is discussed in Chapter 4.

Table 6.3: Industrial Estates in Coastal Maharashtra, 1993-94

District	*Number of industrial estates*	*Industrial area acquired (ha)*	*Industrial plots developed*	*Industrial plots allotted*
Raigad	4	2832	1966	1386
Ratnagiri	4	1405	1470	1300
Sindhudurg	1	275	400	314

Source: Statistical Abstract of Maharashtra State, 1993-94 and 1994-95, Directorate of Economics & Statistics, Government of Maharashtra, Mumbai, 2000.

Ratnagiri District

Maharashtra Industrial Development Corporation (MIDC) developed large industrial areas adjoining Ratnagiri and Chiplun, where basic infrastructure facilities like land, water, electricity are provided to the entrepreneurs at concessional rates. The Government gives 20 to 30 per cent of grant and other concessions over the investment made in this district for new industries. Concessions in sales tax are also given on the sale of manufactured products. Ratnagiri district has emerged as a hub of chemical industries as half of the industries in this district belong to this category. In addition, there are pharmaceutical and dyes manufacturing units, plastic, cement, and a variety of other industries. Only one-fourth of the industries are of large scale and the rest are classed as medium. An overwhelmingly large number of industries (27 out of 36) are located at MIDC's industrial area developed at Lote Parashuram. Riverside industries, Indo Deutch Metallurgical equipment, were established at Lote in 1989. Kesar Petroproducts and Rupal Chemicals were established here in 1991 and 1994. Ms. Konkan Synthetics, Unit No. 2 was established at Karanjari, in Dapoli tahsil in 1994. The rest of the industries are located at MIDC's industrial estates near Ratnagiri and Chiplun. The Enron's gas based thermal power station near Dabhol in Ratnagiri district became operational in 1995. J.K. Files, Narmada Cement, Henkel, NOCIL and Finolex are the major chemical industries. Finolex plant at Ranpar-Golap in Ratnagiri tahsil manufactures resin-powder necessary to make plastic. Among the 289 working factories, 29 employed 50 to 99 workers (18 per cent of total workers), 18 factories employed 100 to 499 workers (30 per cent of workers). There were only five factories employing more than 500 workers (25 per cent of district's workers) in 1999. In order to provide skilled manpower to the newly established industries, Government Polytechnics are established at each tahsil. There are two Government Industrial Training Institutes at Ratnagiri and one at Chiplun. New Industrial Training Institutes were opened at Guhagar, Mandangad, Khed, Sangameshwar, Rajapur since 1996-97 and at Dapoli and Lanja since 1997-98. Several recognized professional courses are taught here. These developments are instrumental in arresting the out-migration of educated youth from this district towards Mumbai.

Sindhudurg District

The industrial estates are developed at Kudal, Kankavli and Majgao (Sawantwadi). The industries in Kudal industrial estate are WG. Forge and allied industries, Sunder Industries, Radha Paper Mills, Konkan Gas. At Kankavli Rural Industries Centre soap and paper making and tanning industries are developed. Making of wooden toys is a cottage industry around Sawantwadi. Cashew processing units are located at Vengurla, Malwan and Kudal tahsils. The cashew in this district has a good national market. A special drink from cashew is manufactured at South Konkan Distillary at Insuli (Sawantwadi). Usha Ispat, a modern large scale project is being constructed at a remote place like Redi with an investment of Rs. 120 crores. It will manufacture 1.74 lakh tonnes of pig iron per year. It is likely to provide indirect employment to about 15,000 people. A steel plant is also coming up at Satarda. Commercial plantation of rubber in Konkan for which subsidy is announced holds good prospects. Vengurla port is developed as a fishing harbour where repair-cum-construction yard for barges and fishing trawlers will be constructed. Fishing industry and horticulture products in the district enjoy a very wide scope for development. There is tremendous scope for the establishment of ice factories, chilling plants, fruit canning industry.

Industries in Goa

Goa has emerged as the focus of industrial activity only after its liberation from the Portugese rule in 1964. At the time of CPCB survey (1997) Goa had 9 large scale and 12 medium scale industries. The largest number of industries (8) belong to the category of foodstuff, including sugar and breweries. Other industries are chemicals, pharmaceuticals, petrochemicals, fertilizer, shipyard, mineral processing and electrical equipments. Ponda tahsil in North Goa district has the largest number of industries (8). Industries are also located in coastal towns of Vasco, Mormugao, Zuarinagar and in Kundain industrial estates developed on Panaji-Ponda road.

Industries in Coastal Karnataka

The Karnataka Government has identified the West Coast as a primary area for investment. Despite this region's rich resource base and strategic location, poor infrastructure conditions serve as a major constraint to economic growth. The state has already exploited the steep descent of rivers originating from the Western Ghats to generate electricity, and additional power plants are proposed. Uttar Kannada is industrially backward compared to other districts in the region. At the time of survey it had only 3 large industries and 16 medium industries. One large and one medium industry was yet to be cleared and one large unit was closed. The industries are based on locally available resources and raw materials. Tiles is the most widespread type of industry with 6 units followed by general engineering (4 units) and foodstuff (3 units). Other industries are paper, mining, iron and steel, chemicals, cement, industrial gases. The industries are located along the coastal belt. Honnavar has the largest number of industries (5) followed by Karwar (3). Dandeli, Bhatkal, Murudeshwar have two units each.

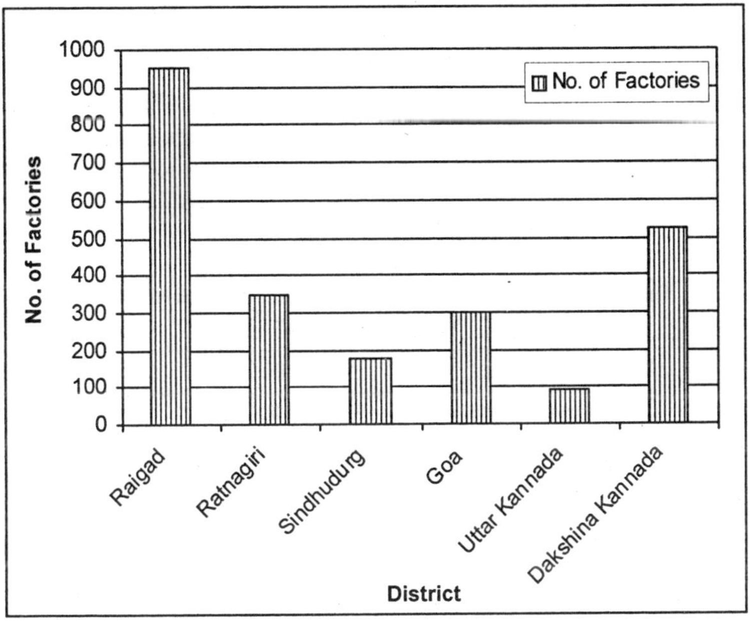

Fig. 6.1: No. of Registered Factories

Dakshina Kannada district is next in importance to Raigad district in terms of industrial development. It has 11 large and 45 medium industries. One large project was under clearance, one was closed and three were yet to be commissioned at the time of the survey. Existing mega industries in Dakshina Kannada district include the Mangalore Refineries and Petrochemicals, BASF Chemicals, LPG bottling plant, Mangalore Chemicals and Fertilizers and the Kudremukh Iron Ore Company. Foodstuff including distilleries, dairy, sugar, edible oil, chocolates is the largest group of industries in this district with one-third of total units, followed by general engineering (9) and rubber products (5). Resource based industries in this district include granite and minerals, stone exporting, timber, wood, plywood and allied products. Other specialized industries are the industrial gases including oxygen and explosives units (two each), plastics. Mangalore urban centre, is a priority area for infrastructure investments to promote sustainable urban, social and economic development. Mangalore and its adjoining area is the focus of industrial development with half of the districts total units. New Mangalore port has also attracted large number of industries. Other industrial centers are Udupi (9 units), Sullya (5) and Manipal (4 units). Baikampady and Yeyyadi industrial areas are developed around Mangalore. Shivally industrial area is developed around Udupi. Udupi district was formerly a part of Dakshina Kannada district. As on 31-3-1999 it had one large scale industry, 11 medium scale units and 5632 small scale units registered with District Industries Centre employing 43,263 workers. Among the factories 12 belonged to Engineering, 8 chemical and 2 textiles. Half (50%) of the small scale

Table 6.4: No. of Registered Factories and Average Number of Workers Employed

Year	*Raigad*		*Ratnagiri*		*Sindhudurg*	
	Factories	*Workers*	*Factories*	*Workers*	*Factories*	*Workers*
1989	194	14,383	172	6,004	149	2,815
1994	487	36,961	261	9,373	162	3,485
1995	514	36,436	262	9,582	162	3,318
1999	953	35,956	338	11,275	—	—
Goa	1995-96		1996-97		—	—
	270	16,395	295	17,519	—	—
	1998	—	—	—	—	—
Uttar Kannada	77	13,457	—	—	—	—
Dakshina Kannada	539	44,839	—	—	—	—

Source: Census of India, 1991 District Census Handbooks of above districts.

Statistical Abstract of Maharashtra State, 1993-94 and 1994-95. Socio-economic Review and District Statistical Abstracts of Raigad, Ratnagiri and Sindhudurg, 1999-2000. Directorate of Economics and Statistics, Government of Maharashtra, Mumbai, 2000.

Statistical Handbook of Goa, 1998-99 and 1999-2000, Directorate of Planning, Statistics and Evaluation, Panaji, Goa.

Udupi District at a glance, 1998-99, District Statistical Officer, Udupi.

Statistical Abstracts of Uttar and Dakshina Kannada districts, 1995.

units in the district belonged to three sectors: food and intoxicants (34%), mechanical engineering and wood (8.5% each).

Figures 6.1 and 6.2 graphically depict the variations in number of factories and factory workers at district level. The leading positions of Raigad and Dakshina Kannada districts and the industrial backwardness of Sindhudurg and Uttar Kannada districts is evident from the diagrams.

Industrial Development after the Commissioning of Konkan Railway

A total of 86 major projects are completed in this region after the commissioning of Konkan railway. It is remarkable that the largest number of projects are in infrastructure sector. Some of them are power projects requiring huge capital investment. About half of the completed projects are in Raigad district, followed by Goa and Dakshina Kannada.

Table 6.5: Major projects completed along the Western Coast, January 1998 to April 2003

Type of industry	*Districts Raigad*		*Ratnagiri & Sindhudurg*		*Goa*		*Uttar Kannada*		*Dakshina Kannada*	
	Comp.	*Stall*	*Comp.*	*Stall*	*Comp.*	*Stall*	*Comp.*	*Stall*	*Comp.*	*Stall*
Infrastructure	11	2	1	5	2	—	3	—	1	1
Petrochemical	1	2	—	—	—	—	—	—	2	1
Chemical	7	1	2	—	2	—	—	—	2	—
Hotel/Resort	1	—	—	—	3	—	—	—	—	—
Cement/tiles	3	1	1	1	—	—	—	—	—	—
Metallurgical	4	—	—	—	1	1	—	1	1	—
ComputerSoftware	7	—	—	—	6	—	—	—	2	—
Drugs & pharmace	3	—	—	—	7	—	—	—	—	—
Miscellaneous	4	—	—	—	8	—	—	—	1	—
Total	41	5	4	6	29	1	3	1	9	2

Source: Monthly Review of Indian Economy, CAPEX, January 1998 to April 2003. CMIE, Mumbai.

Coastal Maharashtra

Raigad

In Raigad district power projects completed were Tata hydroelectric power stage I at Bhivpuri, Jindal Steel's captive power project and

New Koyna transmission line project by MSEB. Rs 500 crore Dolvi power project and gas based 110 mw project to be managed by NTPC are proposed. JNPT in Raigad district is a major port of the country. Here three port development projects were completed for handling liquid cargo.

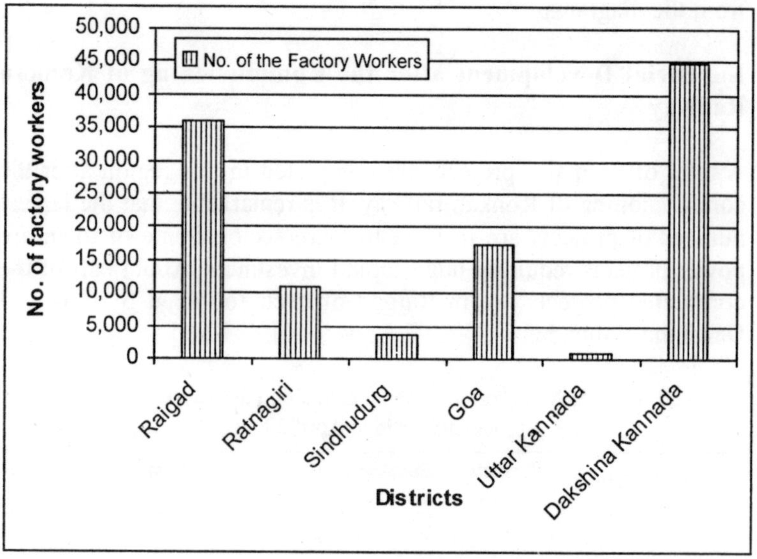

Fig. 6.2: No. of Factroy Workers

A container terminal was set up by converting the 4 existing bulk berths on BOT basis. Bids are invited to prepare report on deepening and widening of Mumbai harbour and JNP channel. A dry dock project was completed at Alibag. Doubling of Diva Panvel railway line was completed by Central railway. Road development projects completed were Taloja creek bridge and Taloja junction flyover bridge. LPG plant with a capital investment of Rs. 320 crores was commissioned by GAIL at Usar in November 1998. It processes 50 lakh cubic metres of natural gas per day received from ONGC which enable to fill up about one crore cylinders per day. Chemical industries completed in Raigad district include acetic acid, nitric acid and Ammonia, naphtha feed supplement and lubricants project. A letter of intent was issued in February 2002 for substantial expansion of a chemical industry located at MIDC, Mahad. Three pharmaceutical projects completed are healthcare formulations at Taloja, Unichem laboratory's drug project at Roha, and Pearl Organics bulk drug project at Patalganga. Raigad district also attracted investment in computer software mainly at Navi Mumbai and Vashi. Call centre, HR centre, Internet data centre, International data centre, computer software and Millenium software park projects were completed here. The metallurgical projects completed include sponge iron and mini blast furnace projects by Jindal group, hot metals and coal washery projects. Other related projects are Patalganga slag cement, readymix concrete project and vetrified tiles at Karjat. Miscellaneous category includes readymade garments and dairy project at Navi Mumbai, chicken processing plant at Taloja and Apollo Hospital's resort cum health retreat at Panvel. Stalled infrastructure projects include Patalganga power project, New Mumbai airport project. IPCL's LLDPE project, Naphtha cracker project at Nagothane and HOC's project at Rasayani. Proposed Raigad Megacity project was shelved in September 2000.

Ratnagiri

Four projects were completed and 6 stalled. Those completed include Dabhol power project, two chemical and pesticides industries and dry plaster project. Stalled projects include Dabhol power project phase II and Finolex cables captive power project, Dabhol LNG terminal and LNG pipeline project, slag cement project at Ratnagiri and Sindhudurg airport project at Parule in Sindhudurg district. IOC is planning to lay a product pipeline parallel to the 1000 km long pipeline being set up by Metropolitan Gas from Dabhol. A series of oil and gas projects with Dabhol as the nodal point will be set up (CMIE, CAPEX, October 2000). In January 2002 bids were invited by Maharashtra government to improve Dapoli Khed road in Ratnagiri district. Bids are invited by Maharashtra government to construct Konal hydroelectric power project at Dodamarg taluka in Sindhudurg district at a cost of Rs. 20.48 crores. Units I and II are to be commissioned by 2004. The number of large and medium industries in Ratnagiri and Sindhudurg districts increased from 10 and 27 in 1995 to 19 and 84 by the end of 2001.

Goa

Goa attracted diversified and specialized industries after the completion of Konkan railway. The largest group consists of pharmaceutical industries, including drug formulations, oral powders by world's leading companies like German Remedies, Hoechst, ophthalmic and injectables, inhalers, bifocal and other lenses, Kodak colour cardiography and X-ray film, Ciba speciality chemicals, insecticides, surface mount technology plant, communication equipments. Six projects are completed in computer software sector, including, software technology park at Dona Paula. Three projects are completed in hotel and resort sector. They are Marriot International, Taj Exotica and Mahindra Holidays & Resorts. Projects in infrastructure sector include power project at Sancoale and Mapusa substation to Kolhapur transmission line project completed by Power Grid Corporation of India. Miscellaneous industries include Finolex cables, optic fibre cables, polyinsulated jelly filled cables, softdrink bottling, granules and palm oil project, pig iron-Mandovi pellets, and slag cement. LPG bottling plant was completed by Bharat Petroleum Corporation at Verna. A letter of intent was issued to manufacture plastic closures, printed/unprinted. Metastrip Copper alloy project at Sancoale was stalled in April 2000. The government of Goa is also planning a thermal power station. It has proposed to declare the port town of Mormugao as special economic zone to attract new industries to the region. Over 30 sq. km of land is earmarked for this project. A 100 bed general hospital is constructed by Mormugao Port Trust. A joint venture of Indian Oiltanking and Zuari Industries plan to set up a petro products terminal at Zuari Nagar in Goa. The terminal would be linked to Mormugao port by a 14 km long pipeline for transportation of petro products. Mormugaon port offshore break water jetty (Goa) project was stalled.

Coastal Karnataka

Uttar Kannada

An industrially backward district has attracted three, i.e. the lowest number of projects after the completion of Konkan railway. All of them are in power sector. A 440 mw Rs. 2896 crores Kaiga Nuclear Power Project started prior to 1994 using natural uranium as fuel. It attained criticality in September 1999, but was completed in November 2000. Karnataka Power Corporation completed Kadra dam and Kodasalli Hydel Project on Kali river in August 1999. A super thermal power station is planned at Nandipur at Mulki, and at Madhabad near Karwar. Kudremukh Iron Ore Company's coke oven project at Karwar was stalled in December 2002.

Dakshina Kannada

Nine projects completed in this district include 2 each in petrochemical, chcmical and computer software sector and one each in infrastructure, metallurgical and miscellaneous sector. Mangalore is the focus industrial activity projects completed at and around Mangalore include Mangalore Refinery and Petrochemical Phase I, LPG bottling plant, Mangalore Chemicals and Fertilizers Urea project and Di-ammonium phosphate plant, Kudremukh pig iron project, computer software and call centre project, amusement and leisure park, Mangalore Tannirbavi barge mounted power project. Stalled projects include NTPC Mangalore power project and Naphtha cracker project at Udupi. Karnataka government plans to develop an export promotion industrial park at Mangalore over 200 acre plot area near Mangalore port. Mangalore Refineries and Petrochemicals is planning to expand its capacity of 9 million tonnes. Additional investment of Rs. 3000 crores is envisaged to raise the capacity to 12 million tonnes. Stalled infrastructure projects during this period include, Mangalore LNG terminal and Mysore-Mangalore-Tellichery rail project. On completion of these projects western coast will have better infrastructure facilities.

Table 6.6: Consumption of Electricity for Industrial Use

(Figs.in' 000 Kwh)

District	1990-91	1999-2000	Per cent of total consumption		Per cent increase
			1990-91	1999-2000	
Raigad	1631,943	2550,392	91.46	89.96	56.18
Ratnagiri	92,250	274,340	55.84	54.19	197.5
Sindhudurg	17,213	23,098	18.65	19.30	34.17
Goa	*1997-98*	*1998-99*	*1997-98*	*1998-99*	—
	627,299	659,480	64.20	64.78	9.22

Source: Maharashtra State Electricity Board, Mumbai. District Statistical Abstracts of respective districts.

Statistical Handbook of Goa, 1998-99 and 2000, Directorate of Planning, Statistics and Evaluation, Panaji, Goa.

During the last 10 years, consumption of electricity has increased throughout the region. With the establishment of new industries, consumption of electricity for industrial use has increased at a higher rate. In Raigad being the most industrialized district 90 per cent of the districts total consumption of electricity is for industrial use. Sindhudurg, being industrially backward, consumes only one-fifth of its total electricity for industrial purposes. The consumption of electricity for industrial use has increased over three-fold in Ratnagiri district.

Foreign Direct Investment after the Commissioning of Konkan Railway

During this age of globalized business environment infrastructure performance is a key in attracting foreign direct investment. The biggest investment opportunities in India today are arising in the infrastructure sector. After the adoption of economic liberalization policy by Government of India in 1991 the foreign direct investment (FDI) continued to flow into India at a healthy pace. The issues such as size, composition and determinants of capital flows into the region have stimulated considerable interest among scholars. The ability to attract FDI is regarded as a prestige issue among Indian states and an indicator of development. Following the trend of economic liberalisation the industrial atmosphere in several states is conducive to set up and attract more industries. The industrial policy of various states in India aims at accelerating and strengthening industrial development and attracting more capital investment with special emphasis on expansion and strengthening of infrastructure and to involve private sector in infrastructure development. Special facilities are provided to encourage 100 per cent export oriented units and those based on sophisticated technology and NRI investment.

Though there are significant geographical differences in India and the pattern of FDI reveals remarkable variations, studies of FDI at regional level are limited. Recent trends in FDI on the western coast of India and the adjoining areas of Konkan Railway after its completion are analysed below. The construction of Konkan Railway commenced in 1990, i.e. slightly before the economic liberalization and was completed and commissioned in January 1998 when some of the concrete results of it in the form of FDI were visible. This study is based on the data published by Indian Investment Centre, New Delhi and CMIE. From the monthly newsletters of Indian Investment Centre published since January 1998, lists of FDI proposals approved by Government of India and foreign technical collaboration proposals approved by Reserve Bank of India were compiled. Monthly review of investment projects in India (CMIE) provided a list of completed projects during this period. FDI proposals on the western coast (statewise) are grouped into sectors for analysis.

The trends in total investment by states over the past decade and the survey of investment projects (CMIE) reveal that Maharashtra ranks first in terms of total investment in ongoing and proposed projects. A closer look at the investment indicates that investments are concentrated in a narrow strip of districts running along the western coast of India. The coastal region of each state has a distinct pattern of investment in various sectors.

Recent trends in spatial distribution of FDI on the western coast are examined on the background of the following emerging contradictory tendencies:

(1) To gain access to local markets is the dominant factor inducing foreign companies to invest in economies in transition. It implies that investors will seek locations within host countries that minimize the access cost to local market, typically in or adjacent to capital or core region. With market access being the main inducement, a substantial portion of all investors can be expected to locate in the capital (or main city) or in proximity to it. Here the largest number of other firms can be found, the largest number of potential customers with the highest purchasing power can be reached, and the infrastructure is most advanced. As a result aggregate costs of investment, transaction and transportation will be the lowest and agglomeration effects may accrue.

(2) Foreign investors do not necessarily select the capital or core region, but rather choose to locate outside the capital where the production costs are lower. Major foreign investments are typically located outside capital cities and economic core regions. Multi-national companies are attracted to new industrial sites where labour is cheap, abundant and disciplined. The economies in transition are seen as repositories of cheap labour. Areas with advantageous structure of wages and pool of appropriately qualified labour are favoured. Manufacturing capacity would be re located from core to the periphery. As a consequence. FDI reinforces the processes of regional fragmentation within host countries. Transportation costs are interpreted as important to locational choices made.

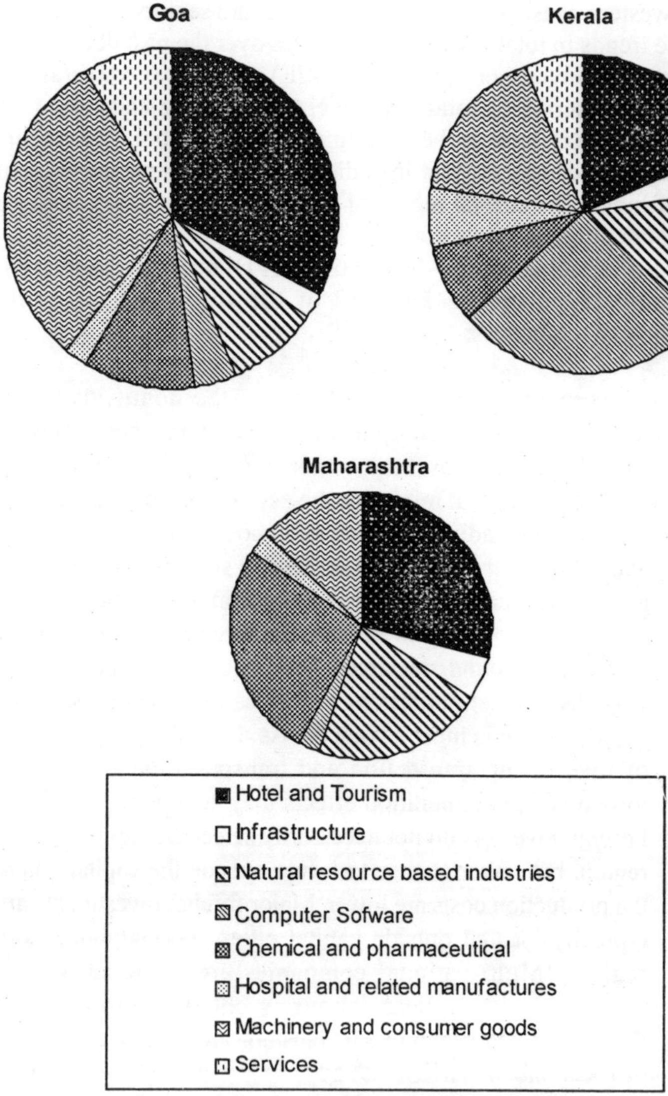

Fig. 6.3: Sectoral Composition of FDI by States

(3) In locating the collaboration units, the foreign countries adhere to normal locational practices usually followed by

> foreign capital to select locations in host countries. They largely preferred the port enclaves and metropolitan cities developed by the colonial power. Western region was the preferred location of collaborations from Britain due to the old colonial economic ties. The western region emerged as the most coveted location for foreign collaboration from the beginning of the Planning era. The European non-British capital that had a late entrance in Indian soil, selected western region as their favoured location.

FDI, specially the Western investment is seen as providing vital capital and technology to our country. It is evident that given the relatively liberal policy framework there is a potential to attract substantial volumes of FDI in technology intensive industries and infrastructure in India. The variety of sectors towards which FDI is attracted towards western coast is simply amazing. This can be seen from Fig. 6.3. Hotel and tourism appears to be the leader among them, followed by Infrastructure and manufacturing (particularly consumer goods) sectors.

FDI in Hotel and Tourism Sector

A significant feature of the FDI attracted towards the western coast after the commissioning of Konkan Railway is the investment in the hotel and tourism sector. This has been discussed in the chapter on Tourism. As per government's guidelines foreign investment to the extent of 100 per cent is permitted in hotel and tourism related services. The statistics of investment in each project reveal that amongst all proposals three are mega-projects with an investment of over Rs. 1000 million in each. Among the 15 countries investing in hotel and tourism sector on the western coast 13 can be called as Western countries and only two—Singapore and Malaysia are the "eastern" countries. USA is the leading country having 6 proposals in this sector followed by Germany (5). France, Mauritius and UK have three proposals each. Projects with the collaboration of USA, Mauritius and Switzerland are spread over the three states while all the five German collaboration projects all with 100 per cent foreign equity. are concentrated in Goa.

Infrastructure Sector

To sustain overall economic growth over a long term requires corresponding growth in infrastructure. Poor infrastructure results in delayed consignments at increased prices and the loss of market. India's industries have been unable to cash on global market opportunities due to infrastructure bottlenecks. Infrastructure in India has consistently lagged behind both in terms of demand supply ratio and quality. According to World Development Report 1994 on "Infrastructure for Development" published by the World Bank, developing countries invest 4 per cent of their national output and a fifth of their total investment in new infrastructure. The result has been a dramatic increase in infrastructure services like transport and power. The Indian government accords priority to attracting FDI in infrastructure along with the export sector as compared to other sectors. The Government is taking several steps to strengthen the infrastructure base in each state and hence after the economic liberalization, the largest share of investment is devoted towards the same. Private participation in developing infrastructure is encouraged under the new policy. 30th Survey of investment projects in India in July 2002 reveals that investment in electricity and infrastructure services have posted a positive growth.

In literature the impact of public investment and physical infrastructure on private investment behaviour has been found to be highly significant and positive. Better infrastructure enhances higher economic growth, which in turn, can provide more resources for investment in infrastructure.

FDI in infrastructure sectors in this region has significantly increased after the commencement of construction of Konkan Railway Several other infrastructure projects were taken up. FDI rushed in for a variety of infrastructure projects in areas benefited by Konkan railway due to improved access. Maharashtra has a major investment proposal of Rs. 1300 mn. by Mauritius for development of Container terminal facilities at Nhava Sheva port. Another proposal is for LNG infrastructure associated with power plants and industrial customers put forth by Enron Co. of USA. This company also completed a power project at Dabhol in Ratnagiri district in Maharashtra in May 1999 with an investment of Rs. 3128 mn.

In Goa the FDI proposals in infrastructure sector include the one for providing world class shipping and logistic services to Indian customers with the collaboration of Panama. A power project was completed at San Coale (Goa) in July 1999. Karnataka coast has attracted two major proposals from Mauritius—one is for setting up a barge and mounted power plant of 170 mw. It involves 100 per cent foreign equity with an investment of Rs. 2565 mn. in Dakshina Kannada district. Another one involves an investment of Rs. 3300 mn. for setting up a coke oven-cum-regeneration power plant based on imported coal in Uttar Kannada district.

Natural Resource Based Industries

This sector includes agriculture, forest and mineral resources. India has adopted a policy of encouraging investments including FDI in food processing, other agro based industries and industries with backward linkages to agriculture. The western coast is rich in natural resources and there is tremendous scope to establish resource based industries with FDI. Due to importance of agriculture in the economy of the west coast, a thrust on attracting FDI into agro processing sector is only natural.

A special feature of FDI in Goa is the utilization of locally available natural resources through these proposals. Goa has also attracted a US. Proposal to manufacture bio stabilizer for agricultural use.

Raigad district has attracted NRI's investment in the field of vegetables, fruits and flowers, tissue culture and rose cut flowers and Israely investment for a unit of cut flowers. Letters of intent are issued to manufacture modern range of modular furniture (two units) and a new unit to manufacture wooden furniture and its components.

An integrated project combining pest control spraying operations, horticulture, soil conservation and soil testing is proposed with NRI collaboration. This district has also attracted FDI for units of meat processing (Netherlands), and granulated chicken manure (Australia).

Mineral resource extracting and mineral based industries display a distinct pattern on the west coast. Goa is rich in mineral resources and it has attracted FDI proposals related to minerals and metals—copper and nickel alloy (Spain), alluminium structural dividers (Israel), metal casting components (Italy), machinery for construction and mining (Germany), gypsum board and accessories (UK). Japan has extended collaboration to establish a unit to process granite and other stone materials and decorative crafted stone and other handicraft stone items at New Mangalore with 100 per cent foreign equity.

Computer Software

In India a vast educational infrastructure in this sector has encouraged that the human resource availability is by itself an important incentive for FDI. High revenue potential of the software sector makes investment in this sector highly lucrative. In Goa there are two proposals—one for developing software for banking application by Switzerland with 99.99 per cent foreign equity and another for compiling, processing and storing of mass data with German collaboration (100 per cent foreign equity). In Maharashtra a software development complex is developed at Mumbai.

Chemical and Pharmaceutical Industries

Chemicals has traditionally been a strong area for FDI though in the past it was restricted largely to technical collaboration. The recent trends indicate the desire of global chemical firms in directly exploiting the domestic market. These industrial units are heavily concentrated in coastal Maharashtra, mainly in Raigad district to the south of Greater Mumbai. Four proposals in this sector are in the collaboration with the UK, two with Germany, and one each with USA, Australia, Czech Republic and NRI. (One is with 100 per cent foreign equity). In addition, there are proposals for expansion of one chemical and two petrochemical industrial units in Raigad district. A letter of intent is issued to manufacture basic oils suitable in manufacturing lubricating oils and greases in Ratnagiri district.

Though Goa has not attracted investment in chemical industry as such, there appears to be a concentration of manufacturing glass, glass products and special purpose lenses in Goa. Two such proposals are with the collaboration of USA and France. There are three proposals to manufacture glass fibre pressure vessels with the collaboration of Malaysia, Mauritius and USA.

Hospital and Related Manufacturing

Another specialized sector which has attracted FDI on western coast is the health sector. In Raigad district a unit for manufacturing adhesive gauze, adhesive tapes, medicinal bandages, lint would come up with NRI collaboration. Another specialized unit in Goa would manufacture lenses used in prevention and eradication of blindness caused by cataract with French collaboration.

Manufacturing and Consumer Goods

This sector holds great attraction for foreign manufacturers. The large size and resource base of India encourages indigenous manufacture rather than imports to service the domestic market. Consumer goods is globally an important sector for FDI. In developing countries like India with a protected consumer goods industry there is a large potential market for technologically superior and branded consumer goods. Consumer durables is one segment which in India has fulfilled its potential for attracting large volumes of FDI despite restrictions. In Goa a variety of consumer goods industries include wrist watches (Mauritius), marker sketch pens, bakery products, pastries and sauses (Netherlands), upholstery and furniture of plastic (Italy), furniture items (Sweden). In Raigad district FDI in consumer goods is limited to two modular furniture units (Netherlands) and a unit of knitted cotton socks (Italy).

Goa has attracted substantial FDI in units manufacturing machinery and miscellaneous products. They include accumulators (Germany), industrial stainers, filters and valves (UK), cellular polyethylene and polypropylene sheets (South Korea), metal casting components for auto industry (Italy), pumps and spare parts of pumps (Germany), machinery-construction and mining (Germany), suspension shock absorber (US) colloidal gold conjugated for use in rapid test (US), partitioning and storage system, alluminium structure divider (Israel), miscellaneous manufacturing (Taiwan). Raigad district would have units of pre-fabricated building material and flat rolled products (Netherland), air pollution control system (US), casting, forging motor engines (UK), hydraulic system for tanks, and cement (NRI).

Service Sector

This sector for long remained outside the priority of the Indian government in the FDI liberalization policies. FDI in tourism related services is included in the hotel and tourism sector. Goa has been the focus of FDI in service sector. Here Panama would provide world class shipping and logistic services to Indian customers, software for banking applications would be provided by Switzerland, consultancy in payphone services by Spain, architectural financial services in construction activities by Germany (100 per cent foreign equity), recruitment and training of personnel and to market sell pressure vessels by USA, compiling, processing and storing mass data by Germany.

The collaborating countries are making adjustments with the new infrastructure arrangements in the host country. USA and UK are perceived as having a superior technology base in India. The share of UK reflects the historical linkages with India. Malaysia and Israel occupy premium positions in investment. Mauritius, which is known as a tax shelter, is also an important source country for India. The trend indicates the foreign entrepreneurs' perception of India as a lucrative market opening up. Foreign entrepreneurs, particularly multinational enterprises are endorsing this potential through their interest in investing in these industries.

Conclusion

Like the completion of Channel Tunnel in Britain, which attracted huge investment in priority transport infrastructure projects, the Konkan railway has played an important role in attracting investment for industrial development in this region. Industrial growth has accelerated in the districts through which the Konkan Railway is passing and also in the hinterlands having connectivity with the rail and roads to the ports along the west coast. The account of industrial situation prior to construction of Konkan Railway and the analysis of projects completed after the completion of Konkan railway support this generalization. A large number of major projects are completed in this region after the commissioning of Konkan Railway.

This reveals that infrastructure development has generated a positive impact on industrial investments. It is remarkable that the largest number of projects are in infrastructure sector. After the commencement of construction of Konkan Railway, several other infrastructure projects were also undertaken in the west coast region. With this, the region has emerged as a focal point of FDI in India.

Mumbai is India's commercial capital and gateway of tourists. Goa is the most favoured tourist destination and a major port. The concentration of FDI over here support the hypothesis put forth at the outset. The investors are seeking locations in or adjacent to capital or core region and port enclaves in order to gain access to local market. The spatial distribution of FDI in various sectors and at locations spread throughout the western coastal region support the second hypothesis as well. Foreign investors do not necessarily select the capital or core region but rather choose to locate outside the capital where production cost is generally less. Transport and other infrastructure (Konkan Railway in this case), is playing a crucial role in their choice of location. The emergence of Raigad district as a hub of manufacturing activities and the opening up of Goa for consumer goods and other manufacturing activities, in addition to tourism, support this generalization. Resource based and primary processing ventures display a distinct geographical pattern on the coast. Not only the regions which were the focus of industrial development benefited by construction of Konkan Railway, but the major beneficiaries are the industrially backward regions of South Konkan and Coastal Karnataka. The Indian Railways and Konkan Railway are working out details for a business plan covering seven industrial hubs along the western coast (*Nav Hind Times*, Goa, 17th January, 2006). This would include an urgent expansion of line capacity along with improvisation so as to break long blocks. These industrial clusters are located in Maharashtra and Karnataka. Indeed, the emergence of Konkan Railway has proved to be a new asset and an instrument to attract FDI and industrial development on the western coast.

REFERENCES

Banerjee Guha, S., (1998). Foreign collaboration in Indian manufacturing during early years of planning (1951-1970): A spatio-temporal analysis, *Geographical Review of India*, Vol. 60(4), pp. 457-479.

Iqbal, B.A. (1998). *F.D.I. flows to India*, Ajanta Publications, New Delhi.

Murphy, A.B. (1992). Western Investment in East-Central Europe: emerging patterns and implications for state stability, *Professional Geographer*, 44(3), pp. 249-259

Ranade, P.S. (1996). Liberalisation and emerging trends of industrialization in Maharashtra, *Yojana*, 40(3), pp. 16-18.

Thomas, Jibby, (2000). Foreign direct investment in India and Pakistan: A comparative study, Ph.D. Thesis, Jawaharlal Nehru University, New Delhi.

Foreign Investment Policy of Government of India, October 1995.

Sjoberg, O. and M. Josefson, (August 1996). The geography of foreign direct investment in Central and Eastern Europe: A model, Presented in *28th International Geographical Congress*, The Hague.

Ranade, P.S. (August 2000). Konkan railway: A new asset for tourism development and environmental improvement in Western India, Presented in *29th International Geographical Congress*, Seoul.

7
Impact on Ports

A port is a key component or a vital system of infrastructure, indispensable to the smooth operations of international trade. Efficient, economical and expeditious infrastructure is a necessary pre-condition for the growth of exports. In its role as a transfer point between land and sea transport networks, a port serves as a pivotal element in the circulation of trade flows. Ports reflect the country's commitment to trade and earnings that stem from trade. Ports serve as a funnel through which the impulses of modernization and development are channeled along with imports. A port is a visible manifestation of economic activity, a symbol of integrated growth centre and an agent of balanced regional development. Establishing a strategic network of ports is essential to promote regional economic development. The up-gradation and expansion of ports is vital to strengthen India's position in world trade and to handle the growing volume of international trade. Commissioning of Konkan railway in January 1998 was a major step towards infrastructure development in India. Konkan railway touches most of the important ports of Maharashtra, Goa and Karnataka. Prior to it due to lack of connectivity, some port towns remained aloof from each other. Konkan railway is a unique railway system in the world because of its length and alignment running more or less parallel to the coast, due to which it has connected several major ports and influenced their volume and composition of traffic. Similar situation does not prevail in other parts of the world. Completion of Konkan railway has provided an impetus to the development of coastal areas by connecting ports or providing connectivity to ports on the western coast.

The Indian subcontinent is blessed with a long navigable coastline with 12 major ports and several minor ports. The 2965 km long Western coast line extending from Kandla to Kanyakumari has 6 major ports, 17 intermediate and 127 minor ports. This chapter aims to study the impact of commissioning of Konkan railway and other measures of infrastructure development on major ports of Western coast which are connected by Konkan railway. These ports are: JNPT (Nhava Sheva in Maharashtra), Mormugao (in Goa) and New Mangalore (in Karnataka state). Ratnagiri in Maharashtra and Karwar in Karnataka, though not classed as major ports, are connected by Konkan railway and hence taken into consideration. The transport connections of these ports with other parts of the country, infrastructure facilities available at these ports and traffic handled by them are analysed. The impact of infrastructure development and up-gradation of infrastructure facilities is visible in the increase in volume of traffic handled by these ports. This is substantiated by comparing the statistics of traffic handled by these ports prior to commissioning of Konkan railway with their present (2001-02) traffic. The changes in the volume and composition of traffic of these ports as a consequence of their increased accessibility and the share of these ports in India's trade are analysed. The unique position of these ports in the national context is highlighted. Mumbai port is India's premier port. It was anticipated that the new Konkan railway line may relieve congestion of Mumbai port by enabling the diversion of some of the ocean traffic to other Western coastal ports. It is attempted to find whether this objective has been achieved.

Inland Transport Connections of Major Ports

Rail linkages with all ports need to be strengthened particularly when new port capacities are being created. All the major ports in India have broad gauge rail connectivity and are expected to be connected through four lane roads with the Golden Quadrilateral Highway Project. The port connectivity project envisages strengthening of the existing routes and laying tracks along new alignment. This involves development of 37 km stretch of National Highway 17 at New Mangalore port and 30 km of NH 4 B and 16 km of NH 54 at JNPT by the end of 2005. There are plans to develop multi modal corridors to facilitate running of double stack containers.

Jawaharlal Nehru Port (Nhava Sheva)

This port was planned with a view to decongesting Mumbai and serving as a hub port for this region. This all weather tidal port is strategically located on the mainline maritime route. There is a common harbour channel for JNPT and Mumbai port. Like Mumbai, it has an advantageous geographical location, excellent infrastructure facilities developed in and around the port, very good railroad connections to the hinterland. The port is well connected to Western and Southern railway. Construction of Konkan railway has increased its connectivity with the southern parts of India, in addition to its earlier connection with Mumbai and other parts of India. A branch of broad gauge railway line connects Panvel to Uran and Nhava Sheva. Doubling of Panvel-Jasai rail link has also increased its accessibility. A highway parallel to railway line connects it to National Highway 17 at Panvel. The port has transport links with Mumbai, Vashi and Uran. The nearest airport is Sahar International Airport of Mumbai. A new airport project for Navi Mumbai is proposed at Panvel. The primary mode of container cargo movement is through road, which accounts for 65 per cent of the total movement. About 33 per cent of the total container cargo movement is through railways.

Mormugao Port

Situated at the mouth of river Zuari, Mormugao port in Goa is approximately 370 km south of Mumbai (Bombay) and 575 km north of Kochi (Cochin). It has a natural harbour protected by breakwater. Mormugao port is connected with the rest of the country through two national highways—NH 17 and NH 4A. The NH 4A eventually joins with NH 4 which connects Mumbai and Bangalore. NH 17A connects Mormugao port to NH 17 at Cortalim. A newly constructed 18 km byepass 4 lane highway (NH 17B) connecting Mormugao port to NH 17 at Verna junction by skirting the Vosco town has considerably eased the restriction to road approach. The inland waterways through the river Zuari and Mandovi are extensively used for transporting ore through barges. The domestic airlines connect Goa to Mumbai, Delhi, Bangalore, Cochin, Pune directly and other major cities through them. The century old metre gauge railway system connecting Mormugao harbour with Madgao has been recently converted into broad gauge. The port railway system is thus connected to South Central railway through which it is also linked to the Konkan railway.

Both these railways together facilitate easy access to the port from any part of the country through the vast network of broad gauge railway system.

New Mangalore Port

Along the Karnataka coastline there are 21 minor ports, a medium-sized harbour at Karwar, a fishing harbour at Malpe and a major deep sea harbour at Mangalore. Mangalore is called as the "Gateway of Karnataka" as it is the only major port of Karnataka. This artificial lagoon port with open berth is situated at the mouth of the river Netravathi, midway between Mormugao and Kochi (Cochin). It is 170 Nautical miles south of Mormugao and 191 Nautical miles north of Kochi port. The port is well connected by broad gauge railway line with Mangalore, places in Kerala and Chennai. Mangalore was connected by Southern Railway's Arsikere Hasan metre gauge line to the hinterland, which was converted as a broad gauge line in 1998-99. The conversion of Mangalore-Bangalore metre guage rail link to broad gauge is currently underway and likely to be completed by the end of 2003. The new railway line could result in shift of a portion of the containerized cargo movement from Bangalore to Chennai and Cochin to New Mangalore port. It had no rail link with other coastal towns of Karnataka prior to construction of Konkan railway. After the commissioning of Konkan railway, Mangalore is directly linked to Mumbai via a shorter rail route and travel time between Gujarat and Rajasthan and this Southern port is considerably reduced. National Highway 17 passes just outside the port, thus making the port accessible to major cities. The nearest airport Bajpe (Mangalore airport) is just 18 km from the port. There are daily flights from Mangalore to Mumbai, Bangalore and Chennai.

Infrastructure Facilities at Major Ports

Ports being a strategic asset, the primary responsibility for development and management of ports in India rests with Central Government. To cope up with the increase in the volume of trade and to face the growing challenges and emerging needs of the 21st century, infrastructure facilities at the ports have been expanded and up-graded. With globalization of the Indian economy, creation and up-gradation of port and shipping services to international standards is essential. There is an imperative need to provide a port system, which should be flexible and dynamically responsible to the trends in the country's maritime trade. Considering the new trends and technologies of the global trading systems, the ports are obliged to meet the ship operators requirements with all sorts of modern facilities and services in order to retain a competitive advantage. The quality of services has become a major factor affecting the customers choice of terminals and ports in the past two decades. The experience on the eastern coast of India reveals that with the increasing industrial development activities and to meet the export demand, the railways were expanded at Visakhapatnam. A study of the West Coast Transport Network (Modak, 1980) had pointed out that the port infrastructure and other backup facilities made available to coastal shipping industry at ports may be considered as less than satisfactory or extremely inadequate, and whatever facilities are provided are hardly utilized according to expectations. The situation has changed after the construction of Konkan railway. Most Indian ports are operating at more than 100 per cent capacity utilization and yet are inefficient when compared to other ports in the world. The following analysis examines how the infrastructure facilities at these ports have been up-graded to cope up with the growing and diversified traffic.

Jawaharlal Nehru Port Trust (Nhava Sheva)

The country's youngest and most modern port was commissioned for commercial operations on 26 May 1989. It was initially planned to be a "satellite port" to Mumbai, mainly with a view to decongest traffic at Mumbai port. The Port is emerging as one of the hi-tech ports of the country. The port has computerized operations with single window operation system. The port was initially planned and constructed for handling only dry bulk cargo and containerized cargo. Responding to the needs of the time, the port facilities have been upgraded for handling a variety of export and import items. This port has got certain advantages which attracted more and more traffic over the period of last 10 years.

- *Port Railway System*: There are two tracks of covered railway platform RMGC 4 rail track and railway siding to facilitate bagging and dispatch operations. Specialized infrastructure is developed for receiving and handling vehicles arriving by train. There are berths for ro-ro vessels for export/import of vehicles and exclusive yards for vehicles.

- *Storage Facilities for Liquid Cargo*: There are separate berths for handling liquid cargo. They are connected by pipelines. These are owned by Indian Oil Corporation and Hindustan Petroleum Corporation who handle POL products, Zuari Industries who deal with phosphoric acid, Indian Molasses Co. Ltd. who handle furnace oil, caustic soda, molasses, etc.
- *Container Terminals*: Considering the growth rate of container traffic, it was planned in 1994 for expanding the container handling facilities. A new Nhava Sheva international container terminal was built with Australian collaboration on BOT basis by converting the 4 existing bulk berths. It started operations since April 1999. Projects are completed for handling liquid cargo. There are two import container freight stations (CFS) and three export CFS. Few more CFS are coming up in the vicinity of the port. Currently the JNPT is the only port that can receive mother vessels up to 4000 TEUs. The port has allotted 1,57,000 sq. m paved area to five different consortium on lease rental basis for managing handling of export containers. One Buffer yard has been developed for storage of export containers till completion of documentation. Future development plans include dredging for deepening of the approach channel, conversion of the existing bulk terminal into a third container terminal, modernization of 3 jetties, construction of second liquid chemical berth, replacement of wharf crane and harbour tugs. In May 2002 the Shipping Ministry approved the Rs. 700 crore bulk terminal redevelopment project at JNP. It envisages conversion of existing bulk handling terminal into a container terminal.
- *Marine Training Institutes*: There is a training ship TS Chanakya, a marine academy and Ocean education and research centre at Navi Mumbai, close to JNPT. JNPT has its own township.

Mormugao Port

Mormugao port was originally commissioned in 1888 by the Portugese rulers of Goa. It was declared as a major port in 1964. It is an open type harbour protected by a breakwater and a mole running parallel to the quay. The harbour is also protected from the south west monsoons as it has been constructed on the leeward side of Mormugao headland. The harbour provides anchorage to about 20 ships during the fair season (October to May). About 8 ships can be accommodated inside the breakwater even during the monsoon period.

- *Power supply*: Reliance Salgaoncar Power Company commissioned its power plant in Goa in 1999 and supplies power to the port. This ensures uninterrupted quality power to the mechanical ore handling plant. Steps are taken to enhance the efficiency and improve the loading capacity of mechanical ore handling plant.
- *Railway system:* Railway lines have been provided to the operational areas including sheds, warehouses and open plots. The broad gauge railway system of the port serves the general cargo berths. A wagon handling system to handle about 2 to 3 million tonnes of iron ore is installed. Wagon loading in broad gauge started from November 1997. A new R& D yard was commissioned in 1999.
- *Bulk liquid discharging facilities*: Specialized facilities are available for handling petroleum products and other liquid cargoes like phosphoric acid, furnace oil, caustic soda, molasses, etc. Liquid cargo is handled through pipelines connected to the tankage provided by private parties. Facilities are available for supply of furnace oil, light and high speed diesel oil to vessels visiting the port at the berths.
- *Ship repair facilities:* A ship repairing complex with dry docking facilities has been set up at the port by Ms. Western India Shipyards Ltd. Besides this, facilities for afloat repairs and others are provided by Goa Shipyards, situated in the vicinity of port area.
- *Construction of Berths:* Construction and operation of two multipurpose bulk cargo berths on BOOT basis for handling about 5 million tonnes of dry bulk cargo primarily coal/coke meant for steel mills being set up in the hinterland of the port by M/s ABG Goa Port Pvt. Ltd. has been completed. The port has planned construction of a fishing jetty at Vasco Bay for the fishing community and modification of the

existing mechanized ore handling facility and acquire new stackers and reclaimers when acquired are expected to increase iron ore handling capacity from 9 million tpa to 12 million tpa.

- *Marine training institute:* There is an Institute of Marine Studies and Seascan Marine Services conducting various courses related to shipping and maintenance.

New Mangalore Port

The Mangalore harbour project was completed in May 1974 and in 1980 it was declared as India's 10th major port. The draft available in the port at the entrance channel is 15.4 metres. It has deep draft berths for handling POL and other liquid cargo. The port is equipped with all types of cargo handling equipments. The port has two jetties for handling crude imports and exports. A new jetty was commissioned in 2002 for the POL products of Mangalore Refineries. A captive coal handling jetty is to be built by Nagarjuna Power Corporation for their proposed 1000 MW coal based power plant nearby.

- *Railway system*: The port is provided with railway sidings connected to the trunk railway with adequate storage space and facilities.
- *Warehousing and liquid storage facilities*: There are covered warehouses owned by Port Trust, Coffee Board and other private agencies, open stack yards and large open storage area near the berths and liquid storage facilities. Oil farm for naphtha and petroleum products are installed by IOC near the oil jetty. Molasses, edible oil tanks are built by Indian Ports Warehousing Co. M/s Mangalore Chemicals and Fertilizers have storage tanks for liquid ammonia and phosphoric acid just outside the port boundary. M/s Universal Agro Exports have storage tanks for storing edible oil. M/s Mangalore Liquid Impex have storage tanks for storing edible oil, cashew shell liquid, etc. The port has separate tank farms for storage of hazardous liquid cargo and LPG M/s L&T have 3 silos for storing cement.
- *Container traffic*: This port has a great potential for container handling as it has the deepest inner harbour at the entrance

channel. There is a regular container service at the port. Container traffic was reintroduced from March 2000.

No marine training institute is located at any port of Karnataka.

Traffic Handled by Major Ports

The sea mode of transport enjoys some inherent advantages in the movement of certain commodities like coal, salt, clinker, tiles, timber and mineral ores over long distances. It is also eminently suited to handle project-oriented traffic.

The three ports together handled over one-fifth (21.18%) of India's total trade in 2000-01.

Jawaharlal Nehru Port

The cargo handled by JNPT is more diversified compared to other ports. The list of principal commodities exported comprises 7 commodities in case of Mormugao, 8 in case of Mangalore port, whereas it comprises of over 50 commodities in case of JNPT. The port handled 18.5 million tonnes of traffic during 2000-01, which was much more than its capacity. Cargo imported (55.4%) exceed the cargo exported (44.6%). The imports comprise POL products claiming the largest chunk of imports (14.3%), edible oils (7.9%), cement (6.8%), fertilizers (2.4%) and other liquids (1.8%). The export list consists of other liquids (4.8%), POL products and iron ore raw (1.6%). Liquid cargo from JNPT comprises Phosphoric/acetic acid, class B & C chemicals, molasses, Mono Ethyl Glycol (MEG), edible oil. Over three-fourth (77%) of cargo handled at this port is containerized, which includes 91 per cent of its imports and 65 per cent of exports. 617 container units were handled by the port for exports and 548 units for imports. JNPT handled over one-third (34%) of the total containers visiting Indian ports in 2000-01 which is the largest among all the ports of India. This proportion was 27.4 per cent in 1999-2000.

Mormugao Port

Mormugao is the premier iron ore exporting port of India. The port accounts for 43.3 per cent of India's iron ore exports and ranks among the top 10 iron ore exporting ports of the world. 81.7 per cent of the port's total traffic consists of exports. Most of the exports are to Japan, South Korea and China. The port handled 19.6 million tonnes of traffic in 2000-01, which was to its full capacity. The bulk of it (93.6%) was overseas traffic and a small amount (6.4 %) was coastal traffic. 92 per cent of total traffic handled by this port consists of dry bulk. The bulk of the export (96%) is of raw iron ore. Iron ore pellets consist of only one per cent of exports. Iron ore mines are located within about 50 km periphery of the port and are operated by private companies. Iron ore is transported in small barges through the Zuari and Mandovi rivers to Mormugao port for eventual loading in large sized vessels. Manganese ore and aluminium ingots are also exported in small quantities. Coke/Coal, POL products, phosphoric acid and fertilisers are the major commodities imported. Coal accounts for 52.5 per cent and POL products 27 per cent of its imports. Coking coal is imported for steel plants at Bhadravati and Hospet in Karnataka. Liquid bulk and container traffic is limited in this port. Only two containers were loaded and two were unloaded during the year 2000-01.

New Mangalore Port

Mangalore is the largest LPG handling port and one of the largest iron ore handling port in India. It is the first port to have state of the art cement terminal. Almost the entire cargo handled in New Mangalore port originates from or destined to coast based industrial units/mines. The port handled 17.9 million tonnes of traffic during 2000-01, which was slightly less than its handling capacity. Over three-fourth (76%) of it was overseas traffic and 24 per cent coastal. A special feature of this port is that it handles the largest proportion of liquid cargo among the three ports. Liquid cargo comprises 63.3 per cent of its total cargo handled. Imports (51.3%) at this port exceed the exports (48.7%). General cargo handled at this port includes granite stones, wooden logs, plywood, pulp, sugar, machinery, etc. Bulk commodities like iron ore concentrate, iron ore pellets, iron fines, iron scrap, sponge iron, copper concentrates, perlite ore, food grains are being handled regularly. The largest traffic handled by this port is of POL products. 69.4 per cent of the imports of this port consist of POL crude and another 2.3 per cent of POL products. LPG constitutes 10.44 per cent of its imports. Fertilisers, cement and food grains are the other commodities imported, comprising 4 to 5 per cent of total imports. Coking coal and edible oil rank next in importance, comprising 2.4 and 1.5 per cent of imports. Over half

(58.7%) of the port's exports are of iron- ore 30.2 per cent as iron pellets and 28.5 per cent as raw ore. Kudremukh Iron Ore Company transports iron ore mined at Kudremukh National Park via 110 km slurry pipeline. The company has its plant within the New Mangalore port for producing iron ore concentrates and pellets for export. POL products constitute 38.7 per cent of its exports, and other liquids 11.4 per cent. Timber and granite are the other export items comprising 3.5 and 2.3 per cent of exports. The containerized cargo moving through the port are coffee, brake drums, cashew, leaf springs, reefer cargo, etc. Traffic at old Mangalore port cater to the needs of Lakshadweep islands.

Ratnagiri Port

Ratnagiri port located in the heartland of Konkan remained neglected mainly because of its lack of connectivity with the railway line. With the construction of Konkan railway, it has emerged on the railway map and is an important railway station. Ratnagiri, classed as minor port functioned mainly as a passenger port catering to the traffic of Mumbai bound passengers which was operating only during the fair season. In 1993-94, it handled passenger traffic to the tune of 4 lakh and in 1994-95 it was 4.78 lakhs. In order to encourage the establishment of industries within the district, various measures were taken by the Government of Maharashtra. It was declared as "economically backward district. Industrial estates were developed. Ratnagiri, the district headquarters and the National Highway 17 passing through this district attracted industries. With the establishment of industries nearby, Ratnagiri was developed as a captive port. M/s. Finolex Industries have constructed their own jetties at Ratnagiri for their import requirements. Ratnagiri port handled 5.21 lakh tonnes of cargo during the year 1998-99. The break up of cargo reveals that it functions mainly as import port. The port exported only 220 m tonnes of cement, while it imported 68,000 tonnes of LPG, 3.16 lakh tonnes of clinker, 1.07 lakh tonnes of ethylene dichloride and 29,000 tonnes of ethylene. Vengurla port nearby is developed as a fishing harbour where repair-cum-construction yard for barges and fishing trawlers will be constructed. The completion of Konkan railway has given impetus for the development of coastal areas. However, it leaves out some of the promising Maharashtra ports like Murud Janjira, Dabhol, Devgad due to inhospitable terrain. They need to be connected to Konkan railway by constructing branch lines in future.

Karwar Port

Traffic surveys of Karwar, Honavar, Coondapur and Malpe ports were carried out by National Council of Applied Economic Research as early as in 1962, but the ports could not develop due to lack of rail connectivity. Karwar was once an important trade centre of the British, French, Portugese and the Arabs. Lack of connectivity by broad gauge railway line with the hinterland in Karnataka hampered its development. Karwar was mainly a service and trading centre. Manufacturing activities are gradually picking up accompanied by rapid population growth. Karwar is classed as an intermediate port. It is now being developed as a naval base of Indian Navy under Operation Sea Bird project. In 1998-99 Karwar handled 2.27 lakh tonnes of traffic, 62 per cent of it consisted of imports. The port traffic increased to 5.48 lakh tonnes in 2000-01. According to Karnataka Chamber of Commerce, there is a very large potential for exports through Karwar for the various products of North Karnataka. Traffic in Karwar port consists of granite, food grains, building materials and generally meant to cater to the needs of Lakshadweep islands.

Increase in Ports Traffic

After economic liberalization, the country has experienced a major boost in its export and import trade. The report of Ministry of Surface Transport reveals India's ports have struggled to keep up with the increase in demand. Total tonnage handled by Indian ports during 1997-98 was 251 million tonnes as against the capacity of 217 million tonnes.

The combined traffic of all the three ports increased from 24.9 million tonnes in 1990-91 to 615.3 million tonnes in 2001-02, recording an increase of 151 per cent within this time span. The increase is most remarkable in case of JNPT as compared to other two ports. The break up of traffic reveals that the imports have increased at a much higher rate (386%) than the exports (60.7%). Among the three ports, Mangalore recorded increase in imports at the highest rate (559%). JNPT is the only port where exports increased over 21 times and imports over 6 times during the last 10 years. The imports and exports at Mormugao port increased at the lowest rate among the three ports (86 and 4% respectively). The values of exports and imports from these ports are compared for the years 1991-92 and

2001-02 in Table 7.1. The table reveals that the value of exports from these ports (Rs. 42,964 crores in 2001-02) is much higher than the value of imports (Rs. 16,817 crores). However, the value of imports in these ports increased at a higher rate compared to exports. The three ports together handled imports worth Rs. 1344 crores in 1991-92, which increased to Rs. 16,817 crores in 2001-2002. The exports from these ports increased from Rs. 5021 crores in 1991-92 to Rs. 42,964 crores in 2001-2002. The imports in 2002-03 further increased to Rs. 23,126 crores and exports to Rs. 62,696 crores. The JNPT recorded the highest rate of increase in exports compared to other ports. The rate of increase recorded at Mormugao is the lowest among the three ports. Import in terms of Rs. from Mormugao have declined in 2002-03 compared to 2001-02, while those of Mangalore have substantially increased within one year. Export values of all the three ports have remarkably increased.

Table 7.1: Value of Imports and Exports from Major Ports

(*Value Rs. Crores*)

Port	*Imports*		*Exports*	
	1991-92	*2001-2002*	*1991-92*	*2001-2002*
JNPT (Nhava Sheva)	1,007	14,466	3,880	41,682
Mormugao	230	1,612	505	552
New Mangalore	106	739	636	730

Source: *Foreign Trade Statistics of India*, Principal commodities and countries, Ministry of Commerce and Industry, Director General of Commercial Intelligence and Statistics, Government of India, Kolkata.

In order to study the impact of commissioning of Konkan railway, traffic handled by major ports in terms of tonnage in the year 1997-98, i.e. the year when the Konkan Railway was commissioned are compared with those in 1993-94, i.e. four years prior to construction of Konkan railway and with those after four years its construction (2001-02) in Table 7.2.

The table reveals that though there is an increase in amount of cargo handled at all the ports, the increase in cargo handled in terms of percentage is higher prior to construction of Konkan railway at Mormugao and New Mangalore ports. Only at JNPT the cargo handled increased at a higher rate after the commissioning of Konkan Railway compared to the earlier period.

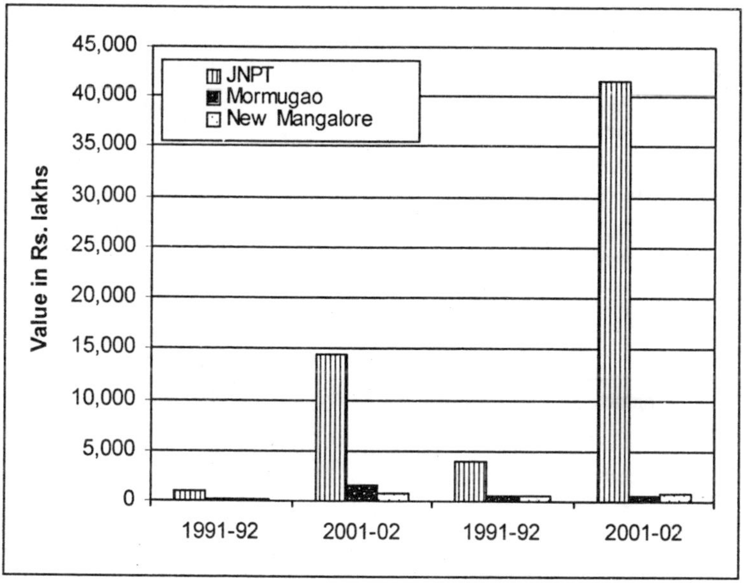

Fig. 7.1: Imports and Exports through Major Ports

Table 7.2: Increase in Cargo Handled by Major Ports

(Cargo handled in 000 tonnes, Increase in %)

Port	*1993-94*	*1997-98*	*2001-02**	*1993-94 to 1997-98*	*1997-98 to 2001-02*
JNPT	3010	8895	24090	162.2	170.8
Mormugao	15570	21182	22930	13.2	8.2
New Mangalore	7090	15282	17510	76.6	14.6

* India Port Report: Ten years of reforms and challenges ahead, I-maritime Consultancy Pvt. Ltd., Navi Mumbai, September, 2003.

Source: Basic ports statistics of India, 1998 99, Traffic handled at Indian ports, Transport Research Wing, Ministry of Surface Transport, Government of India, New Delhi.

Changes in the Composition of Cargo by Broad Categories

India's major ports have been designed for the previous generation of general cargo ships of up to a maximum of 60,000 tonnes and up to 10-15 metres draught and to handle specific categories of cargo, which have declined with time while other types of cargo gained importance. Apart from increase in the value and volume of traffic, the composition of cargo handled by these ports has undergone changes. Traffic handled by these ports in terms of four broad categories of cargo prior to commissioning of Konkan railway and in 2000-01, are shown below in Table 7.3.

Table 7.3: Changes in Composition of Cargo at Major Ports

(PC share of category of cargo to total cargo at each port)

Port	Dry Bulk		Liquid bulk		Break bulk		Container	
	1995-96	*2000-01*	*1995-96*	*2000-01*	*1995-96*	*2000-01*	*1995-96*	*2000-01*
New Mangalore	78.77	33.43	14.75	63.35	6.47	3.01	0	0
Mormugao	87.6	92.28	11.93	6.63	0.33	0.86	0.22	0.1
JNPT	27.26	7.03	2.03	15.93	11.19	0.16	59.2	76.87
Total of 3 ports	73.11	45.28	10.66	27.8	4.15	1.34	12.08	25.53
Share of 3 ports to all India	29.88	22.37	3.65	13	8.74	4.95	23.2	44.14

Source: *The Economic Times Ports of India*, 1999 and 2002, Bennett Coleman & Co. Mumbai.

About half (45%) of total traffic handled by these ports in 2000-01 consists of dry bulk, over one-fourth (28%) of liquid bulk and the remaining one-fourth of containers. Traffic in liquid bulk handled by the 3 ports together increased over three fold (333%) and container traffic increased by 250 per cent within this time span. These figures clearly reflect the impact of infrastructure development. Handling of liquid bulk and containers at ports requires specialized facilities. The ports could achieve this growth only due to up-gradation of infrastructure facilities. Traffic in dry bulk remains almost stagnant (increase by 2.6). Break bulk traffic at these ports has declined. The analysis of traffic at each port reveals that dry bulk traffic (Mechanical) of Mangalore and JNPT has declined, whereas dry bulk conventional traffic at these ports has increased. However, conventional traffic constitutes only 17 per cent of dry bulk traffic. Traffic in liquid bulk cargo recorded most remarkable increase in case of JNPT and Mangalore ports, whereas its traffic at Mormugao port has declined. Container traffic is confined to JNPT. The increase in share of trade of 3 ports in liquid cargo and containers in India's trade reveals the impact of infrastructure development. The liquid bulk and container traffic could increase only due to specialised infrastructure facilities developed at these ports.

Changes in the Composition of Cargo at Major Ports

With infrastructure development, the import list of these ports has become more diversified. More commodities can be imported after the completion of specialized infrastructure facilities for their imports. Exports of POL products from New Mangalore port after the commissioning of Mangalore refineries has increased the share of these ports in India's exports. Share of exports of other liquids from these ports has also increased.

Jawaharlal Nehru Port

The trade handled by this port is more diversified at present compared to earlier years. There is an addition of commodities in the import list, like POL products, edible oil, which were not handled at JNPT a few years back. Quantity of imports of cement and other liquids in this port has increased, and that of fertilizers has declined. The volume of exports of iron ore and food grains has decreased, whereas exports of POL products and other liquids have increased. The share of dry bulk handled by this port has significantly declined due to increase in liquid bulk cargo. The share of container traffic has expanded due to improvement in infrastructure facilities for their handling. It is only due to JNPT that the 3 ports handle 44 per cent of India's container traffic.

Mormugao Port

Being a predominantly iron ore export port, dry bulk constitutes over 90 per cent of its trade. The share of dry bulk handled at this port has further increased over the period because of up gradation of facilities for its export. Unlike other two ports, there is no export of POL products from Mormugao port. This port also experienced decline in import of POL products and other liquids, mainly because of the diversion of trade of these commodities to other ports.. Though iron ore is the major export item of this port, iron ore and pellets are also found in its import list of 2000-01. Another item in the import list, which is not found in the lists of other two ports, is iron and steel. This is due to the establishment of a steel mill nearby. There is an increase in import of coal and fertilizers. Exports of iron ore, other liquid and coal from this port have increased. In 1999-2000 imports in Mormugao port were mainly from South Africa, Bahrain and Australia. Half of the exports from this port were sent to Japan, and the rest to China and South Korea.

New Mangalore Port

The composition of traffic from Mangalore port has also undergone changes. From a predominantly dry bulk handling port in the past, it has now emerged as a major liquid bulk handling port. Composition of exports was more diversified in the nineties. Coffee, Manganese ore, chrome ore, iron fines, plywood, molasses were included in the export list upto 1995-96, and do not figure in the list thereafter. Exports of POL products and other liquids have increased, while those of iron ore pellets, granite and timber have declined. The tonnage of commodities imported like scrap, sugar, phosphoric acid, liquid ammonia has decreased gradually from 1991 to 1995-96. The imports of POL crude and products, edible oil, fertilizers, chemicals, coal, food grains, cement have substantially increased. LPG was not in the import list up to 1995-96, and added thereafter with the completion of infrastructure facilities for its imports.

Share of three Western Ports in India's Trade

The share of the three western ports in India's export trade is much more than in its import trade. This is most remarkable in case of export of POL products, other liquids and iron ore—raw and pellets. Over half of India's iron ore exports, over one third of the country's exports of POL products and one-fourth of the exports of other liquids were handled by these 3 ports in 2000-01. POL products are also imported in substantial quantities. Cement is the major import item, comprising 81 per cent of India's imports, followed by food grains. Imports of other liquids and fertilizer raw materials in these ports comprise about one-third of India's imports of these commodities. Edible oil is another important import item. These ports make a significant contribution in India's container traffic, wherein export trade claims a larger share than the import trade.

The three ports together handled 15.23 per cent of India's total trade prior to construction of Konkan railway (in 1995-96). This share has increased to 21.18 per cent after the completion of Konkan railway (in 2000-01). Changes in the share of country's imports and exports over the period are shown below in Table 7.4.

Table 7.4: Share of Traffic at Major Western Ports to all Indian Ports

Year	*Import*	*Export*
1995-96	7.13	32.18
2000-01	14.24	32.09
Per cent increase, 1995-96-2000-01		
Average 3 ports	186.76	32.5
All Indian ports	43.7	32.8

Source: Compiled from *Economic Times Ports of India*, 1999 and 2002.

It is remarkable that imports at the three ports have significantly increased compared to the Indian average growth of imports during this period. The growth of exports at these ports has kept pace with the Indian average. Therefore, the share of imports at the 3 ports has doubled and that of exports remained the same. The infrastructure facilities created at these ports have facilitated the imports of certain commodities and the increased accessibility of these ports has facilitated their distribution in other parts of the country. The changes in the share of these three ports in India's export and import trade of principal commodities is shown in Table 7.5.

Table 7.5: Traffic Handled by Three Ports as per cent of India's Foreign Trade

Commodity	*Share of India's imports, 1995-96*	*Share of India's imports 2000-01*	*Share of India's exports 1995-96*	*Share of India's exports 2000-01*
POL (Crude)	0.42	11.01	—	—
POL (Products)	7.18	14.26	—	36.61
Edible oil	4.27	26.06	—	—
Other liquid	13.07	33.02	11.76	24.88
Iron ore, raw, pellet	92.87	—	63.97	52.93
Fertilizers	7.61	18.76	—	—
Fertilizer Raw material	5.08	7.37	—	—
Food grains	—	41.21	2.37	—
Coal	1.77	5.83	—	
Iron and Steel	—	2.51	—	—
Cement	—	81.56	—	—
LPG	—	n.a.		—
Total	7.13	14.27	32.5	32.00
Container	21.30	42.43	23.36	46.10

Source: Compiled from *The Economic Times Ports of India*, 1999 and 2002.

Ranks of West Coast Ports in the Indian Context

It is interesting to study the changes in the ranks of West coast ports among the 12 major ports of India as a consequence of increase/decrease in volume of trade at these ports as a consequence of their infrastructure development. The ranks of Mumbai port are specially given for comparison. Congestion was a major problem at Mumbai port resulting in a long turnaround time. With the availability of an 'alternate', namely JNPT, and up-gradation of infrastructure facilities at other western ports, this problem has been eased. JNPT has emerged as a competitor to Mumbai port, drawing on its traffic volume. In 1997-98, the average turnaround time at New Mangalore port (4.1 days) was the lowest among the three ports. At JNPT and Mormugao it was 5.6 days. At Mumbai it was 8.3 days. The average of all Indian major ports was 6.3. Average turn round time of all ports came down to 4.1 days in 2000-01 and 3.7 days in 2001-02. It was reported less than 3 days for New Mangalore and JNPT. Average pre-berthing detention time in days was also the highest for Mumbai (3.6) compared to Mangalore (1.8), Mormugao (1.4) and JNPT (2.8). These four ports provided employment to 43 per cent of India's port workers in 1998.

Table 7.6: Ranks of Western Ports among Indian Ports

Port	*Overseas trade*		*Coastal trade*		*Imports*		*Exports*	
	1995-96	*2000-01*	*1995-96*	*2000-01*	*1995-96*	*2000-01*	*1995-96*	*2000-01*
Mumbai	3	4	2	4	4	5	1	5
JNPT	7	7	12	10	8	7	9	7
Mangalore	8	8	10	9	12	9	6	6
Mormugao	5	5	11	11	11	11	2	2

Source: Compiled from statistics published in *Economic Times Ports of India*, 1999 and 2001.

Mumbai port caters to 9.6 per cent of the country's sea borne trade handled by major ports.

Table 7.6 reveals that Mumbai port has lost its ranks in every respect within this time span. From No. 1 position in export trade in 1995-96, Mumbai has slipped to fifth position in 2000-01. Mumbai, first ranking port in handling container trade in 1995-96, now ranks third. JNPT is now India's first ranking port in container trade. It occupied No. 2 position earlier. JNPT has improved its ranks both in terms of export and import trade. Though its rank in overseas trade remains the same, its rank in handling coastal trade has also improved. Mangalore port has improved its ranks in terms of coastal trade and import trade, whereas Mormugao port has retained its ranks in all respects over the years. Though ranks low in imports, Mormugao has retained its second position in India as an export port.

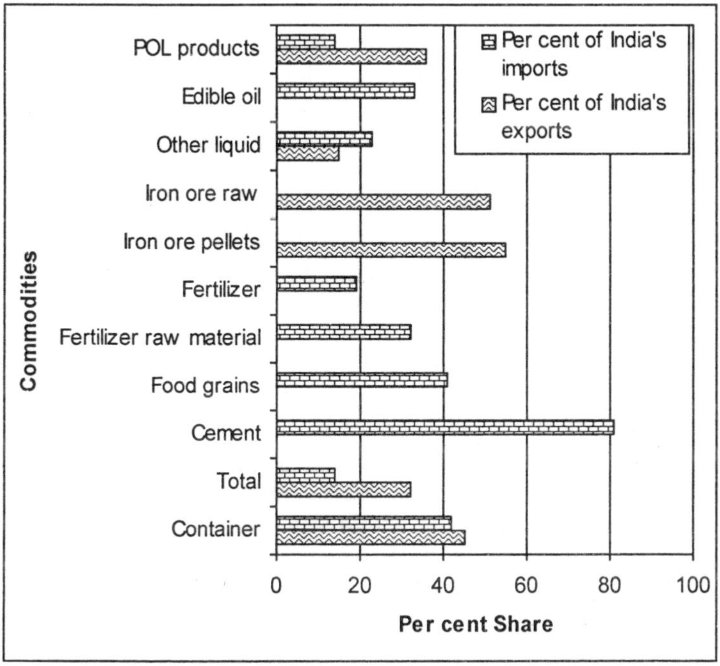

Fig. 7.2: Traffic Handled by 3 Ports as per cent of India's Foreign Trade, 2000-01

Government Policies and Plans for Future Development

Government policies play a major role in the performance of ports. The Government adopted a policy of providing broad gauge rail connections to all major ports and connecting them with National Highways. The port connectivity project is expected to have significant impact on the Indian ports sector. The Government has decided to initiate the process of phased corporatization of major ports to enable them to operate in a market oriented economy with adequate flexibility. Ports selected for this process in the initial stage are JNPT, New Mangalore, Mormugao and Tuticorin. Government of Goa has proposed to declare the port town of Mormugao as special economic zone to attract new industries to the region. Over 30 sq. km of land is earmarked for this project. A joint venture of Indian oiltanking and Zuari Industries plan to set up a petro products terminal at Zuari Nagar in Goa. The terminal would be linked to Mormugao port by a 14 km long pipeline for transportation of petro products. Karnataka government plans to develop an export promotion industrial park over 200 acre plot area near Mangalore port. The Central Government has also planned to develop hub ports, one on the east coast at Chennai and one on the West coast at Navi Mumbai. Union Ministry of Chemicals and Petrochemicals have proposed to set up sea port based chemical and industrial estates in all the 9 coastal states of India. A task force has been set up to work out the modalities. The project would also entail development of exclusive chemical terminals at the sea ports by private sector on a build-own-operate basis (CMIE, CAPEX, August, 2000).

The ports on the West coast have been designed for the cargo. No passenger traffic is handled at the three western ports under consideration except at Mumbai. Under the Konkan Development Programme, Maharashtra Government has planned to renovate 7 new ports and 19 jetties at the cost of Rs. 100 crores. In Raigad and Sindhudurg district 6 jetties are under construction. The recent trend of new industries coming up within this region is to construct their own jetties to handle their project related traffic. Under the Private sector Ispat Industries proposes to expand the capacity of its captive jetty at Geetapuram port in Raigad district at a cost of Rs. 10 crores. Its capacity would be increased from 3.4 million tonnes to 8 million tonnes. 100 acres of land is acquired for storage yard. Licence Agreement was signed for development of Rewas and Dighi ports on 17 March 2002 by Maharashtra Maritime Board. The natural draft at these ports will be increased and 5 berths (2 bulk, 1 petrochemical and 2 containers) will be constructed at Rewas in the first phase. Liquid cargo terminals to carry crude, chemicals and LNG will be constructed at Dighi port. A new jetty project at Dherand in Raigad district by Dherand Chemical Terminal Ltd. is under implementation. Government of Karnataka has approved substantial expansion of Malpe fishing harbour project in Dakshina Kannada district which is likely to be completed by the end of 2002.

Ports in International Perspective

Ports are basically a viable business. Total revenue earned by each of the port under consideration is more than the expenditure. The three ports earned revenue of Rs. 727 crores in 1997-98 compared to the expenditure incurred (Rs. 448 crores). The revenue earned by these ports increased at a higher rate than the expenditure incurred over the last 10 years. The revenue earned by JNPT per tonne (Rs. 329) is much higher than the Indian average (Rs.144 in 1998-99) and the other two ports. While performance matrices have improved considerably, they continue to lag international standards. In todays global internet trade era ports are no longer merely goods handlers, rather they strive to become efficient and comprehensive container logistics centres. Containerisation which brought about a technological revolution in the transport world is still to make an impact in India. JNPT ranks 26th and Mumbai 45 in the list of world's top container ports. Indian ports are costlier than other ports in the region for handling containers. World Bank Report (2000) points out that the total costs of moving a container through a terminal in India are over 70 to 80 per cent higher than those in Japan and US. None of the Indian ports figure in the list of fifteen major container ports in the world (Ha, 2003). Productivity at container terminals in the JNPT is less than half that of Colombo. Port facilities in developing countries, on average move cargo from ship to shore at only 40 per cent the speed of the most efficient ports (World Development Report, 1994). Thus there is tremendous scope for improvements in infrastructure sector. Our country has a long way to go to make its presence felt on the international scene.

Conclusion

The Rakesh Mohan Committee recommended that the major ports urgently need to upgrade their handling technology, modernize their equipment and management and raise adequate resources both for the creation of additional port facilities and to improve existing ones. A lot needs to be done to improve infrastructure facilities at ports of Western coast. Our analysis reveals that infrastructure development at and around the ports of Western coast has a positive impact on the traffic handled by these ports. After the commissioning of Konkan railway, the connectivity of these ports with other parts of the country has increased. The current port connectivity projects under implementation are expected to have significant impact on the Indian ports sector. By reducing logistics cost, a strong positive impact on growth of trade is likely. With the improvement in infrastructure facilities for handling the traffic of specialized items, trade at these ports has become more diversified. JN Port with more diversified composition of cargo recorded higher growth compared to other ports. As a result of infrastructure development and growth of traffic at the three ports, the share of traffic handled by Mumbai port in India's traffic has declined. Our analysis reveals that the improved connectivity of western coastal ports after the construction of Konkan railway has helped to relieve the congestion at the Mumbai port. Most of the west coast traffic which used to converge at Mumbai port is now routed to other ports. Increased routing of traffic to these ports has also led to a better utilization of their infrastructure facilities. With infrastructure development at ports and improvement in goods traffic handled by Konkan railway in due course of time, traffic at these ports will be further augmented.

REFERENCES

Anto Joseph (ed.) *The Economic Times Ports of India,* (2002). Bennet Coleman & Co., Mumbai.

Basic Ports Statistics of India, (1999-2000). Traffic handled at Indian ports, Transport Research Wing, Ministry of Surface Transport, Government of India, New Delhi.

Carol Androde (ed.) *The Economic Times Ports of India,* (1999). Bennett Coleman & Co., Mumbai.

Foreign Trade Statistics of India, Principal commodities and countries, (March 1999 and March 1992). Director General of Commercial Intelligence and Statistics, Ministry of Commerce, Government of India, Calcutta.

Ha, Myung Shin, (2003). A comparison of service quality at major container ports, implications for Korean ports, *Journal of Transport Geography*, 11, pp. 131-137.

India Infrastructure Report 2001, (2000). Issues in Regulated Market Structure. 3i Network, Oxford University Press, New Delhi.

India Infrastructure Report 2002, (2002). Governance issues for commercialization, 3i Network, Oxford University Press, New Delhi.

India Port Report: Ten years of reforms and challenges ahead, (September 2003). I-maritime Consultancy Pvt. Ltd., Navi Mumbai.

India Reducing poverty, accelerating development, (2000). A World Bank Country Report, Oxford.

Infrastructure, (February 2003). Economic Intelligence Service, Centre for Monitoring Indian Economy, CMIE, Mumbai.

Maharashtra, Key Data, (1999-2000). The Blue Book, Sumangal Press, Mumbai.

Modak, S.K. & D.M. Nachane, (June 1980). *West coast Transport Network: An optimal inter modal mix:* a study sponsored by the Government of Maharashtra, University of Bombay.

Monthly review of investment projects, January to December 2002 issues, CAPEX, CMIE, Mumbai.

The India Infrastructure Report, Policy imperatives for growth and welfare. (1996). Expert group on the commercialization of infrastructure projects. Published on behalf of Ministry of Finance, Government of India, by Dr. Rakesh Mohan, NCAER, New Delhi.

8

Environmental and Safety Measures, Resettlement and Employment Generation

Environmental Conservation by Konkan Railway

For ecological conservation it is important that railways should be aligned and their structures designed in a manner which is "eco centric" rather than purely techno centric. Indian railway continues to make efforts to reduce the adverse effects of railway operations on the environment through adaptation of cleaner technologies, energy conservation measures, afforestation of vacant railway land, virtual embargo on the use of wooden sleepers, phasing out of steam locos, etc. In the building of this transport infrastructure, i.e. Konkan Railway, the issue of environmental preservation was of paramount importance. Ecology of the area traversed by Konkan railway is unique. It was, therefore, the constant endeavour of the Konkan railway corporation to cause minimum damage to the ecosystem and the beauty of the environment. Konkan Railway Corporation's policy of ecological conservation was stated as follows:

(1) All structures should aesthetically merge with the beautiful Konkan surroundings.
(2) Construction activities should not degrade the ecology and the environment.

The construction was planned so as to fully preserve the ecology of the area and enhance the beauty of the environment. All the structures (station buildings, etc.) aesthetically merge with the beautiful Konkan surroundings. Its construction activities have not degraded the ecology and the environment. After studying the alignment with respect to various environmental aspects the alignment chosen by KRCL was approved by various State Governments as it was not posing any serious threat to the fragile eco system of the region. Under the Environmental Management Plan (EMP) KRC took a number of eco-friendly measures even before starting construction. Konkan railway line is running almost parallel to the coast line in certain sections. It is maintaining safe distance from the sea, nowhere less than a km and is free from the danger of high tide levels. Extreme care was taken to avoid heritage, religious and community structures during the construction. The railway line passes at a distance of 4-5 km away from the south-western border of Cotigao wild life sanctuary in Goa and hence does not pose any threat to the sanctuary. Due to hilly and mountainous terrain of the western coast cutting of rocks, excavation on the mountain sides, building of bridges and tunnels on such terrain was likely to give rise to geological disturbances. Therefore, the construction of Konkan railway was undertaken with considerable care and caution, so as to minimize the frequency of land slides and hillslips. Other issues like erosion of soil, destabilization of hillside terrain, interruption or disorganization of drainage pattern and flow of streams, possibilities of flash floods and loss of soil nutrients were also taken care of. Perennial springs and the resultant streams are numerous along the Konkan railway line. Adequate cross drainage across the railway embankment has been provided in order to prevent their blocking by earth or rock dumps and other refuse.

The Konkan railway is subjected to very intense rain during the monsoon between June and September. This is responsible for the flooding of river channels, silting of lower reaches of river channels and soil erosion at number of places. An adequate number of culverts are necessary and are constructed for the free flow of water, particularly during heavy monsoon shower. The mass movements like soil creep, rock falls, debris slides and landslides are recorded at various sites. These landslides occur in remote and comparatively inaccessible areas of Sahyadris and its offshoots. The landslides in the various ghats like Kumbharli, Amboli, Varandha, Amba are an annual feature of this region. These landslides are the result of instability of slope created due to the constructional work along the roads. This is coupled with the several deep soil and route cuttings.

Use of Concrete Sleepers

Konkan railway decided as an environment friendly measure, to utilize only concrete sleepers throughout its construction. Wooden sleepers were totally prohibited. This implied preservation of forests. KRCL avoided deforestation by use of cement concrete sleepers. Wooden sleepers were avoided even at traditionally common locations such as bridges, points and crossings, level crossings and switch expansion points. There is no trace of even a single wooden sleeper over a route length of 760 km. Concrete sleepers were used throughout. They were manufactured at four plants located along the track—in Chiplun and Kudal (Maharashtra), Margao (Goa) and Kaikini (Karnataka). The Chiplun plant also produced crossing sleeper sets for the entire project. The factories also produced special concrete sleepers required for switch expansion joints for bridges with extra holes for guard rails and for level crossings with extra holes for check rails. The sleepers were steam-cured till they attained a strength of 400kg/cm^2 and then placed in a water tank for 14 days. They were handled mechanically in the factory and even at the unloading site. The sleepers were transported by road to the site. At places where the formation was ready, they were unloaded directly by crane on the side slopes of the completed banks.

Afforestation Activities/Tree Plantation

Consciousness of the need to protect forest is now widespread among the intelligentsia and the media. Its effects began to be seen in the slowdown in the rate of loss of forest cover in the 1990s. The western coastal region with year round warmth and moisture provides ideal conditions for fast growing tree plantations. In the Environmental Impact Assessment reports of Konkan railway, it was advised to undertake massive tree plantation. An afforestation programme was taken up on embankments and acquired lands. For the construction of railway line, the minimum width of land was acquired. The cutting of trees was restricted to the minimum possible. Turfing along the embankment slopes using local species was carried out as recommended. KRCL embarked upon a massive tree plantation scheme on the strips of land on either side of the railway line. All vegetation that was destroyed by the new Konkan railway line was adequately compensated by new plantation. The Corporation is exploring the possibility for development of surplus land by cultivating medicinal/herbal plants and other plantations (The 10th Annual Report of KRCL, 1999-2000).

As recommended by the MOEF, the most effective species to absorb the various elements of traffic related pollutants are *Azadirachta indica* (Bevu or Neem), Tamarindus indica (Hunsemara or Tamarind), *Ficus benghalensis* (Aladamara or Banyan), Termindalia chebula (Hallalemara or *Arjun*) and *Dalbergia sissoo* (White Beete or *Shisham*). All these species are the major constituents of the existing avenue plantations. These are also the preferred species in the avenue-plantations by the Karnataka Forest Department. In order to reduce noise and air pollution, double rows of trees along the railway track are planted in certain sections as advised, so as to form a screen. Local plant species which provide a number of non wood produce to the local residents are planted. They have enhanced the beauty of the landscape as well as provide shelter and food for animals.

Tree plantation programme in the premises of railway stations and staff colonies started in 1997 when the construction work completed. A variety of ornamental and fruit bearing trees and teakwood were planted during the monsoon season of 2002. Fruit bearing trees include coconut, mango, *chikoo*, etc. Ornamental trees include Feltafunu, *Gulmohor*, *Ashoka*, etc. The roadside plantations undertaken have greatly enhanced the aesthetics of the road and also function as a pollution arrester and prevent surface runoff in stretches prone to soil erosion. No negative impact is visible on forest vegetation and the fauna during operation phase. Over most area it is not land degradation, but land reclamation.

Statistics of tree plantation by major type of species are available for each station of Konkan railway in Karwar region. Tree plantation progress during monsoon 2002 is given below. The figures indicate actual trees planted and survived in the sections of Goa and Karnataka.

Table 8.1: Trees Planted along Konkan Railway

Section	*Boganvilla*	*Acetia Catechu*	*Fruit bearing*	*Ornamental*	*Teak*
Pernem to Balli	25625	35700	384	2511	845
Balli to Kumta	23000	24766	395	1190	4500
Kumta to Shiroor	41000	32671	1112	402	200
Shiroor to HPCT	20000	18153	268	1137	310
Total	118625	111290	2159	5240	5864

Source: Office of Regional Engineer, Konkan Railway Corporation, Madgao, Goa.

Afforestation Activities in Goa

Goa's forest department made consistent efforts for afforestation over the plateau. The outcome is increase in area under forest. According to the then Forest Conservator, the afforestation was being done with the specific purpose of providing villagers in the surrounding areas with small timber, fodder and fruit, thereby reducing pressures on the natural and degraded areas of forest.

Afforestation activities undertaken by Goa's Forest Department since 1990-91and area brought under each activity during 1990-91 to 1998-99 is given below in Table 8.2.

Table 8.2: Afforestation in Goa

Schemes of afforestation	*Area covered in Ha*
Social forestry schemes	689
Fuel wood and fodder	1190
Rehabilitation of degraded forest	4060
Integrated afforestation	500
Soil conservation	450
Western Ghats Development Programme	2840
Conservation of Mangrove ecosystem	953
Minor forest produce+	1066
Compensatory afforestation	333
Other schemes*	651
Total	12732
Seedling distribution in lakhs	202.21lakhs

+ Includes medicinal plants

* Other schemes include Habitat improvement, Rubber, cashew and oilseed plantation, wildlife management, IREP, JRY, NWDP, Tourism

Source: Goa Forest: A Statistical Report, Forest Department, Government of Goa.

The above table reveals that under the afforestation activities undertaken in Goa during the last decade, rehabilitation of degraded forests has covered the largest area of 4060 ha. followed by Western Ghats Development Programme (2840 ha). Fuel wood and fodder rank next in order. Attempts are also made for the conservation of Mangrove ecosystem. Compensatory afforestation is also undertaken in lieu of the forest land diverted to other development projects. During 1998-99 967 ha and during 1999-2000 710 ha. of additional area was brought under forest plantation. There has been extensive planting under the Social Forestry programmes on private, communal and non-forest public lands over the past decade. Separate nurseries are developed to supply seedlings under social forestry schemes. In Dakshina Kannada district as per NRSA data forest plantations were observed over 1187 ha. in 1988-89.

Table 8.3: Afforestation in Raigad District

(1990-91 to 2000-01)

Schemes of afforestation	*Area covered in Ha.*
Plantation on community and private wasteland	899.66
Plantation under Western Ghats Development Programme	755.00
World Bank aided Maharashtra Forestry Project	3971.00
Rural Area Development Programme	553.00
Afforestation for soil and moisture conservation on special sites	836.00
Plantation for coastal shelter belt	61.50
Afforestation under Jawahar Rozgar Yojana	285.63
Afforestation and pasture development under employment guarantee scheme	667.66
Plantation of forest species for industrial and commercial uses	1210.00
Development of minor forest produce	380.00
Massive Afforestation programme	278.00
Restoration of degraded forest	2175.00
Eco development eco preservation	615.00
Compensatory afforestation	2113.00

Source: Data compiled by Social Forestry Division, Raigad-Alibag, Government of Maharashtra.

Under horticulture development programme introduced under wasteland development schemes by Government of Maharashtra, the main thrust in Konkan is on plantation of Alphonso mangoes and cashew and it is expected to bring an area of two lakh hectare under mango and 1.5 lakh ha under cashew by 2000. Coconut plantations are developed along the coast of Sindhudurg district.

A report from Goa (Anthony Simoes, 2002) has severely criticized the project on environmental ground: " Konkan railway project is a prime example of how a project can wreak ecological havoc of monstrous proportions. Ecological disaster is manifesting itself in various forms all along the present alignment. "Inadequate environmental management measures have contributed to significant degradation of the region's valuable natural resources such as fisheries, mangrove forests, beaches, estuaries, and coastal forests. On the contrary, Goa Today (November 2000), a popular news magazine reported that "in spite of overriding objections from environmentalists in Goa, there is no evidence of any major environmental disaster due to the Konkan railway line even after three years of operation."

Operational Safety Measures Taken by Konkan Railway

Konkan Railway line has been constructed through difficult terrain of loose soil and non uniform strata. The railway line has to face all the vagaries of the nature. This coastal region normally receives very heavy rainfall in the monsoon season. The nature's fury affects the rail services also. The Konkan Railway Corporation is taking all precautions to minimize the occurrence of accidents/possibility of disruptions in future. KRC has expedited implementation of safety technologies to make train journey safe on its entire track. These measures include installation of inclinometers, anti-collision devices and high strength steel nets. Safety measures have also been taken to protect the life of passengers in the event of accidents inside tunnels. Fire extinguishers have been installed and sufficient lighting arrangement made in all tunnels.

All man-made structures on Konkan Railway are designed to take care of seismic forces and rainfall intensity. Konkan Railway is the only railway in India having a permanent Geologist for inspecting and monitoring cuttings and tunnels as well as supervising safety works. The system of Konkan Railway is automated for quick responses in times of accidents and follow up action required to be taken. Due to the peculiarities of the terrain, Konkan Railway has built up various safety features like boulder nets and geo-netting for increased safety of running trains and improved signalling. High strength steel nets have helped prevent some major accidents. A focused patrolling system has been designed for greater safety of goods and passengers. Intense short beat patrolling and portable field telephones using the optic fibre based communication system were introduced in the monsoon of 1998. Konkan Railway has started with special safety works as a strategic plan to reduce the probability of any mishap because of landslide or boulder fall. All vulnerable cuttings are progressively stabilized. The embankment side slopes are being stabilized by stone pitching, berming and increased use of ballast on the permanent way. The side slopes in the excavated cuttings are being fenced to prevent rock falls and other collapses finishing upon the railway track. Sand bags are stacked at the mouth of tunnels to act as buffers.

Konkan Railway has built up various safety features like boulder nets and geo-netting for increased safety of running trains and improved signalling. A focused patrolling system has been designed for greater safety of goods and passengers. Permanent staff is maintained for inspecting and monitoring the various tunnels and cuttings. Intensive patrolling at the assessed vulnerable sections is done regularly.

The bridges on Konkan railway have been designed for 100 years return period flood based on the latest hydrological standards. Hence on account of adequacy of the water ways, the bridges on Konkan railway are not vulnerable. The embankments on the Konkan railway have been so designed that they do not exhibit any vulnerability.

Workers are permanently posted at certain places to do 'running repairs' on the embankments to prevent them from sinking. Landslides and rock falls in cuttings are a regular feature, particularly during the monsoon. Tunnels keep on losing chunks of material due to water logging and vibrations of passing trains. KRCL has taken various measures since then to overcome these environmental problems.

Raksha Dhaga, a predictive failure warning device, indigenously designed and developed by Konkan Railway has proved its efficacy. It sets off an alarm when a boulder or a branch of a tree falls on the chord. A total of 200 km *Raksha Dhaga* is being installed and about 800 numbers of devices which work with the *Dhaga* will be installed at all critical cuttings along Konkan Railway route to give advance information to the engine driver with warning light and hooter signal. (Konkan Railway Press Release, 27th July 2003). All rock cuttings above 15 metres will be provided with special Konkan Railway designed high strength steel boulder nets. In all rock cuttings toe protection wall or pitching (putting a layer of rock along the slope) will be provided. For the high strength nets for rock cuttings deeper than 12 metres, 700,000 sq. metres of netting has been done with different design strengths as required for the locations.

Konkan Railway is installing indigenously developed "Inclinometers" which will act as sensors and detect the movement of soil and any boulder falling on the tracks from a rock cutting. About 30,000 inclinometers are being installed in deep soil cuttings before the monsoon of 2004. Wherever full protection of steel nets are not provided, inclinometers are installed and they have prevented some serious accidents. Electronic pendulums having mini radio transmitters linked to a radio modem which in turn communicates with ACD technology based train control system can provide predictive warning so that simultaneous failure of cutting and arrival of train do not take place. Electronic rain gauges capable of measuring intensity of rainfall and coupled with ACD technology network through wireless connectivity is also being developed by KRCL for real time reaction to rainfall. The anti-collision device (ACD) is an innovative micro processor based network of ACDs forming *Raksha Kavach* and is designed and developed by Konkan Railway to prevent all kinds of train collisions. ACD when fitted to trains, stations and at level crossings will prevent collision and will also provide safety to road users at level crossing gates. The Railway Minister launched the national project implementing the anti collision device on 20-1-2004. With the launch of this project Indian railways have taken a giant leap forward in improving safety of train passengers and also of general public at railway crossing gates. It is proposed to cover the entire broad gauge route of 52,600 route km of Indian railway with ACD at an estimated cost of Rs. 1580 crores. (Konkan Railway Bulletin Board, 20 January, 2004).

A village voluntary force—It was proposed to create a village voluntary force for maintenance of the track during the monsoon season. The scheme was as follows: Each village close/adjacent to the Konkan Railway track will identify 5 to 10 (as required) willing men to be given basic training in track men's duties, drain maintenance, turfing embankments in their areas, use of emergency communication and use of tools on the lines of Home Guards, carry competency certificates and includes oath taking to create sense of responsibility and provided uniform when on duty. The honorarium/compensation for the part time emergency employment during monsoon or accident being fixed at Rs. 100 per day. This step will provide proactive links to the villagers and provide supplementary income to some during off-season, develop a sense of belonging to the villagers who will feel they are as much helping safe carriage of millions of lives on the tracks. For the Corporation it is a cost effective move. The proposal is still under consideration and yet to materialize.

The Konkan Railway announced a separate time-table for its trains during the rainy season to ensure safety of passengers while maintaining punctuality in view of the heavy monsoon in Konkan region of Maharashtra and its effect on the hilly terrain along the route. The new time-table will come into effect from June 15 till October 13, 2005. Departure timings of some of the trains has been advanced at originating points and the frequency of some of the trains has been reduced, (Konkan Railway Press Release 2-5-2005).

Accidents on Konkan Railway

In spite of the operational and safety measures stated above and though Konkan Railway trains run at a much lesser average speed against the promised speed of 160 kmph, accidents do take place. On the background of road accidents, number of persons killed and injured in railway accidents is relatively less. As per information received from the Safety Department of Konkan Railway, the total number of accidents since the commissioning of Konkan railway till 31-10-2001 was 24. The break up of the accidents was as follows:

- Accidents due to natural calamities: 13 Nos.
- Accidents due to other causes: 11 Nos.
- *Casualty*—Killed 13, Injured 40, Simple injuries 32 persons and seriously injured 8.

It is evident that number of accidents on Konkan railway route due to natural calamities is more than accidents due to other causes. Natural calamities in this case include boulder fall, land slides and cutting collapse. Boulder fall is the major factor causing maximum (10) accidents. Frequent incidents of boulder fall are expected in the initial years. There were 21 instances of boulder fall and landslides causing track damage and one instance of ballast wash out due to heavy rain. Track was also damaged due to derailments. In addition to derailments due to natural calamities and accidents, there were 11 other derailments mainly goods derailment. Some derailments have occurred in yards. Other causes of accidents were fire, weld failure, rail fracture, track and CW defect, signal defect, failure of loco drivers, failure of pantry car staff, failure of road vehicle driver, mechanical/ material failure. Number of days train services remained suspended during monsoon season due to excessive rainfall, landslides, boulder fall, sinkage/washout of bridges, approaches, etc. are as per details given below:

Monsoon season	*No. of days train services remained suspended*
1998	Nil
1999	9 days
2000	21day

Accidents lead to interruption of traffic. Interruption to traffic in hours since the commissioning of Konkan Railway is as follows:

Year	*1998-99*	*1999-2000*	*2000-01*	*2001-02 (upto 31-10-2001)*
Interruption hrs.	71'51"	91'50"	269'55"	9'25"

Source: Safety Department of KRCL.

In July 1999 a major landslide was experienced causing disruption of traffic. This was due to combined effect of earth tremors in summer months measuring 3.2 to 5.2 on Richter scale and unprecedented concentrated rainfall, three times the average that caused disruption not only on Konkan railway, but also on National Highways and elsewhere. The worst accident of its nature on Konkan railway took place on 22 June, 2003 at 10 p.m. between Vaibhavwadi and Rajapur road in Sindhudurg district. It involved derailment of Train No. 904 Karwar Mumbai Central Holiday Special train. The accident occurred near the mouth of a tunnel during continuous and heavy monsoon downpour. The train hit a boulder that had bounced on to the track due to a landslide in the area and derailed. Once the engine hit the boulder, four coaches went off the track. The coaches were so badly mangled in the impact that 25 passengers were injured and 50 passengers were crushed to death (KR Press Release, 23 June 2003). Even communication became a problem as the optic fibre cable was damaged. Continuous downpour hampered the restoration of the track and train services. Speed restrictions were imposed on Konkan Railway route and some trains remained cancelled or rerouted for a period of one month. In another major accident on 16 June 2004, 14 people were killed. The mishap of boulder rolling over the track occurred a few seconds before the train passed the spot. The mishap occurred at a cutting which was 8 feet high and was considered very safe. Due to incessant rains for two days, there was seepage in the soil which might have loosened the boulders situated a top and rolled them on to the track. The Railway Board expressed serious concern about the increasing number of accidents on railways and has named the Konkan railway zones as one of the dangerous zones within its network. (*Times of India*, November 8, 2003).

The Northern section of Konkan railway, i.e. Maharashtra and Goa are more prone to accidents. The cases of boulder fall and landslides have occurred in Ratnagiri and Sindhudurg districts where the alignment passes through a number of tunnels or nearer the hills, than in the Southern side, i.e. coastal Karnataka, where the alignment is nearer to coast and away from hills. The safety record reveals that Diwan Khavati-Khed, Adavali-Vilavade, Ratnagiri-Nivsar, Karanjadi-Diwan Khavati, Bhoke-Ratnagiri, Rajapur-Vaibhavwadi sections are more prone to accidents due to natural calamities which led to suspension of traffic. A publication in Goa reported that during the first three weeks of the 1998 monsoons, there were at least 13 derailments, 132 landslides and 2 embankment collapses. Margao-Mumbai trains were running late everyday, the delays ranging from 3 hours to 23 hours. Trains were derailed due to stones being washed down onto the track. Most of these occurrences were unreported as they took place in places where people had no access to the media. In the first five months eight people died on the exposed tracks in Goa hit by trains in an area that was not very easily accessible.

Indian railways have published zonewise figures of unmanned railway crossing accidents and number of persons who lost their lives during the last three years—2002 to 2005. An average of 85 such accidents have taken place per year in India wherein an average of 147 persons were killed every year. It is remarkable to note that no such accident has taken place and no casualty on Konkan railway during this period.

The cost of damage due to accidents is on increase every year as shown below:

(Figs. in Rs. lakhs)

	Department	*1998-99*	*1999-2000*	*2000-2001*
1.	P. Way (Railway Track)	24.25	16.25	28.00
2.	Loco	28.68	22.50	0.15
3.	Carriage and Wagons (C&W)	1.00	83.00	122.70
4.	Signal and Telecom (S&T)	—	1.75	6.01
5.	Commercial	—	16.00	—
6.	Electrical	—	1.00	—
	Total	53.93	123.50.	182.86

Source: Safety Department of Konkan Railway Corporation Ltd.

The Konkan railway could become vulnerable to increases in temperature and rainfall and changes in the sea level in 50 years, according to a study conducted by IIM Ahmedabad. IIM took Konkan railway as case study. The study "Climate change impacts on industry, energy and transport" conducted for a joint programme of the Union Environment and Forest Ministry and the UK Department of Environment said temperatures are projected to rise as much as 3 to 4 degress Celsius, rainfall will increase and there will be more sea surges towards the end of the 21st century (*The Telegraph,* September 13, 2005). The report observed that the region through which the Konkan railway passes experiences moderate to heavy rainfall and that marine ecosystems are sensitive to climate changes. The increase in rainfall can cause waterlogging and erosion while sea level changes can lead to flooding. Konkan railway along the coast could be in trouble with rising sea level. The Konkan railway passes through a geologically unstable area. There is heavy rainfall along the tracks and possibility of landslides. The study predicted extreme climatic events increasingly disrupting the Konkan railway services. (*The Times of India*, September 9, 2005) The report said temperature increases over the 21st century can directly affect the stability and strength of the building materials and indirectly increase the need for air-conditioning on train so more repairs and maintenance will be required. The Konkan railway officials say the current design code takes into account the normal climate change variability. 20 per cent of its repairs and maintenance expenses are due to climatic factors like heavy rain (*The Indian Express*, September 9, 2005).

Resettlement and Rehabilitation Measures

Resettlement is a sensitive area involving competing economic, social, and political interests that may be difficult to balance. There should be proper understanding of involuntary resettlement with its social, cultural, psychological, economic and environmental ramifications. Resettlement is now viewed as a development issue. Policy makers accept that inadequate attention to resettlement does not pay in the long run and displacement effects should be minimised. They propose that resettlement and rehabilitation should be made a part of the development process so as to advance the well-being of the oustees. All kinds of resettlement losses need mitigative measures. The three important elements of involuntary resettlement are (i) compensation for lost assets and loss of livelihood and income, (ii) assistance for relocation including provision of relocation sites with appropriate facilities, (iii) assistance for rehabilitation to achieve at least the same level of well-being with the project as without it. In order to meet the World Bank's policy to restore the economic and social base, people losing livelihood need three things: compensation for lost assets and income, transfer and relocation assistance; and help to rehabilitate and restore their lives. Asian Development Bank's policy recognizes all persons affected by the project as eligible for compensation or rehabilitation irrespective of legal or ownership titles. It aims that people affected should be at least as well off after resettlement as they were before. Those displaced would have the first rights to the benefits from the project. Resettlement programmes in general are limited to statutory monetary compensation for land acquired for the project, and occasional development of a resettlement site. Monetary compensation for land alone may not be adequate. There is the risk that affected persons might spend their cash quickly and become impoverished or that women's and children's subsistence needs might not be met if cash compensation is paid to the head of the household. It has been observed that if the members of the community are paid a one time compensation, it can help them survive for a maximum of three years. Only if they are able to find alternative employment would their livelihood be protected. Measures to restore income and living standards can substitute for compensation payments provided that such measures are sufficient to meet the policy objective.

Compensation Paid in Konkan Railway Project

In India the responsibility for planning and implementing resettlement rests with the government and other project sponsors. The successful operation of the construction activities is mainly dependent on the promptness of payment of compensation. Compensation refers to the amount paid under the Land Acquisition Act for private property, structures and other assets required for the project. In this context, compensation refers to payments made by the Government when title of a property is transferred from a private entity to the Government. The compensation for land is fixed under section 23 (2) of LAA, where the amount to be paid is the market value +30 per cent solatium (compulsory). If this amount is paid within one year, then an additional interest of 9 per cent is paid to the beneficiary ; if it exceeds one year, then an additional interest of 12 to 15 per cent is paid. The Konkan railway project involved land acquisition from 42,000 land owners. Most of the land owners willingly came forward to give consent for the acquisition of their land for this project. They gave consent letters much before they received the compensation, so that work could start immediately. In case of Konkan railway, land owners received interest at the rate of 9 per cent from the date of consent to the date of publication of the Section 4 notice, and 12 per cent after the publication of Section 4 till the date of award. Wherever there was any dispute over the ownership or possession of land between the parties concerned, the matter was referred to the District Court under Section 30 for a decision and the compensation amount deposited in the same court. Cases in which people were not satisfied with the compensation were referred to the Special Land Acquisition Officer for further reference to the District Court. Such claims for enhancement of compensation could be made within six weeks, only if the compensation was received under protest. Compensation was also provided for trees and other structures acquired. For this the valuation of trees was done by technically qualified persons. Trees acquired were grouped into two broad categories: fruit trees and trees used mainly as fuel wood. The number, variety and age of the trees was considered. The Zonal Agricultural Officer provided their fruit value and the Deputy Conservator of Forests was consulted for wood value. A mango tree fetched compensation ranging from Rs. 2000 to 10,000. Cashew trees were valued at minimum of Rs. 1000 and jackfruit at Rs. 2000. Compensation was paid for over two lakh trees acquired. The cheques of compensation worth Rs. 144.88 crores were delivered to the people's doorsteps. The compensation paid by KRCL sectorwise is shown below:

Table 8.4: Compensation Paid for Land and other Structures Acquired

	Mahad	*Ratnagiri North*	*Ratnagiri South*	*Kudal*	*Panaji*	*Karwar*	*Udupi*
Land acquired	608	726	882	759	866	1555	819
Structures acquired							
Houses	—	137	300	218	35	365	783
Wells	—	1 5	54				
Other structures	171	86	—	83	330	1100	1402
Compensation paid Rs. Crores*	10.58	8.41	3.92	12.04	18.29	17.69	20.55

* The above cost does not include 30 per cent solatium.

Source: Salelkar, S.V. (ed.) A Treatise on Konkan Railway, Konkan Railway Corporation Ltd., Navi Mumbai, 1999, pp. 90-92.

The structures on the property were evaluated by PWD authorities. Other structures acquired included hutments, plinths of houses, cattle sheds, burial ground, samadhi, schools, temples, mosques, tanks, wells, boaring pumps. The houses situated on the proposed alignment which were to be acquired were surveyed. The names of owners, type of houses to be acquired were collected from the field. The majority of the occupants were either marginal farmers or workers in the category of low income group. The people were also having livestock like cow, buffalo, sheep, poultry, etc. All this was taken into consideration while fixing the compensation amount. As per the Environmental Impact Assessment Report for the section of Goa, number of residential units coming on the alignment was 25. However, valuable agricultural land has been acquired and hence more amount of monetary compensation was paid compared to Mahad and Ratnagiri sectors in Maharashtra. Number of houses and other structures acquired was also more in Coastal Karnataka (1148 houses) compared to Maharashtra (655). This has also resulted in more amount of compensation paid in Karwar and Udupi sectors compared to other sectors.

Proper rehabilitation in case of this project has been beneficial from economic, social and environmental considerations of the region and has contributed to improved project quality and impact. There was no local resistance to the project due to inadequate resettlement. Our field studies point out that land losers are now better off than they were earlier. With proper planning and implementation, the flow of project benefits was visible right from the construction stage.

Employment to Land Losers

There had been a persistent demand that KRC should provide employment to locals and specially those whose land has been acquired for the construction of the railway line. Employment policy of KRC gives preference to such land losers, subject to their eligibility and suitability for the available jobs. The policy, however, does not provide for one job to each land loser family. It needs to be appreciated that the number of land losers being more than 40,000, it would be impossible to provide a job to each land loser family. KRC just does not have employment potential of this order. Most of the staff required for construction phase was recruited on temporary basis from local sources by the seven Chief Engineers at Mahad, Ratnagiri (North), Ratnagiri (South), Kudal, Panaji, Karwar and Udipi. Later, as a matter of policy, for the requirements of the operational staff KRC decided to offer the posts in operation phase to construction phase staff subject to their eligibility. As a general rule, recruitment to posts in Public Sector Undertakings is required to be made by public notifications of vacancies and selecting candidates by open competition on the basis of merit. However, for speeding up the progress of construction, as also to fulfil the aspirations of people of the region, KRCL recruited most of the construction staff in junior capacity from amongst local people all over the jurisdiction. A committee of MPs which inspected the work of KRCL expressed satisfaction at the representation afforded to various states on which the route of KRCL passes.

On completion of the construction of railway line for recruiting the permanent staff for operational phase, KRCL adopted very rational and transparent policies giving preference to the local staff appointed on contract basis for construction and thereafter to give preference to land losers. Staff for operational phase were selected out of available construction staff, trained and posted. Wherever construction staff were not available, new recruitments were made giving preference to land losers. In accordance with the directive from Regional Labour Commissioner (Central) 36 contract workers engaged for cleaning, sweeping etc. were regularized. They were all local people. In case of premature death, a member of family of deceased was given direct employment. 75 such candidates were provided employment on compassionate ground.

In order to compete for the new jobs of Asstt. Station Master, guard, commercial clerk, TTE, Office Asstt. gangman, pointsman, khalasi, all the land losers were given fair opportunity. KRCL issued two notifications in the local press inviting applications exclusively from the land losers, one on 30-5-1996 for 40 posts of probationary ASMs and second on 6-7-1996 for 250 posts of gangmen all over Konkan railway. In response, a large number of applications were received. They were processed. The eligible candidates were subjected to a written test. Those who passed the written test were subjected to psycho-technical test and interview. The successful candidates were trained and appointed as ASMs. All the applications received against these notifications by the Chief Engineers were registered statewise. The percentage of land lost was taken as a rational criterion for deciding priority amongst the applicants. Thus, statewise lists of applicants in the descending order of percentage of land lost were prepared and candidates in that order were called for interview, physical and medical test. Those found suitable were appointed. For balance requirement for full scale operations, it was decided to organize a competitive open market selection to be conducted by KRC instead of placing indents on RRBs. As such, a press notification was issued on 4-10-1997 to fill up posts like Train guards, Asstt. Loco Drivers, ASMs. The notification made it clear that other things being equal, preference would be given to land losers. The final results were on the basis of total marks, with a preference to land losers who qualified. From the status note and other documents issued by KRCL, it is clear that the Corporation has been quite fair to local people and to land losers in the matter of employment. In Goa three recruitments for land losers were made from 1996 onwards and more than 200 people were recruited for jobs reserved for land losers. It is the policy of Konkan railway to reserve all future vacancies for land loser.

Allotment of Stalls at Railway Stations to Land Losers

All the commercial stalls at the stations have been given by preference to land losers and local residents. In Goa more than 80 per cent of the stalls have been given to land losers. This has helped in employment generation. Stalls at railway stations awarded in favour of land losers and local residents between Kolad and Surathkal stations of Konkan railway are shown below:

Table 8.5: Allotment of Stalls at Railway Stations

Units	*Total stalls*	*Awarded in favour of*	
		Land losers	*Local residents*
Tea stalls	43	22	21
Mobile trolleys	14	14	—
Book stalls	19	11	8
Restaurants	4	—	4
Parking contracts	9	—	9
Milk Parlours	15	—	15
General Stores	3	3	—
MTDC/GTDC	3	—	3

Source: Office of the Konkan Railway Corporation, Belapur.

Other Measures of Rehabilitation

In accordance with the directive of the Railway Board normally the railways do not make any provision for rehabilitation of displaced persons on acquisition of their land apart from giving them compensation as awarded by the Special Land Acquisition Officer. On Konkan Railway, however, as a humanitarian measure and also to expedite handing over of land with dwelling houses, all displaced persons were allowed to take away building materials released after their dwellings were dismantled for use in their new houses. Rent for four months, i.e. time required to put up a new house was paid by Konkan railway in addition to all other compensation. About 150 hutments in Madgaon yard, belonging to tenants had to be cleared. The tenants were not due any compensation, but were given alternative land temporarily along with Rs. 2500 for construction of their dwellings. Displaced licensees of South Central Railway in Madgaon were also allotted plots on licence for carrying on their trade.

Employment Generation by Konkan Railway

Konkan Railway Corporation Limited started basically as a temporary organization created specifically for the project of construction of a new railway line. For such a gigantic construction a large number of engineers and supervisors were required. It was difficult to meet this requirement from the Railway system as very few volunteered to work on Konkan railway, which being a Corporation could offer no real incentive. The staff required for this were obtained through three sources:

(1) Borrowing services of regular employees of railways and other government departments/organizations, i.e. deputationists.
(2) Re-employment for short term of capable employees, who had retired recently.
(3) Direct recruitment on temporary contract basis by:
   (i) Corporate office through open selection, and
   (ii) Zonal Chief Engineers from local population, registered with employment exchanges.

Konkan Railway provided immediate employment during construction. For speeding up the progress of construction, as also to fulfil the aspirations of people of the region, KRCL recruited most of the construction staff in junior capacity from amongst local people all over the jurisdiction. A number of retired engineers who were otherwise fit and suitable were engaged. Special recruitment was organized in the corporate office and 53 assistant civil engineers, 52 junior engineers (graduates) and 55 technical assistants (diploma) were recruited after inviting applications from first class degree/ diploma holders on daily basis with regular pay scales being given after six months. They recruited more than 350 degree and diploma holders. These fresh engineering graduates were given a crash course in IRICEN, Pune to handle the assigned job. The project provided direct employment to 2000 officers and staff and indirect employment through contractors to about 25,000 workers. Considering the temporary nature of employment, the staff required for construction phase comprising of 2174 persons was directly recruited on temporary basis from local sources by the seven Chief Engineers at Mahad, Ratnagiri (North), Ratnagiri (South), Kudal, Panaji, Karwar and Udupi, without following the normal procedure of competitive selection up to March 1995. In 5 different categories like, Assistant Station Master 89 staff members were selected by Railway Recruitment Board on all India basis and supplied in response to indents placed by KRCL in terms of instructions of Railway Ministry in 1993.

Another source of employment provided by Konkan railway to local population in rural areas is at manned level crossings. Where the road crosses the railway track at the same level, a level crossing is provided. There are two types of level crossings, one which is manned and which is provided with gates or lifting barriers and the other type is unmanned with only gate posts provided on either side of road and track. At the manned level crossings, the gateman has a duty hut and living quarters. On Konkan Railway all manned level crossings have been provided with telephones connected to nearest stations. Among the 113 level crossings on Konkan Railway of B, C and D class, 80 are manned. Employment is provided to these people and housing to their families at level crossings by Konkan Railway.

Konkan Railway also helped in improving access to employment, particularly non-agricultural-mployment. After its commissioning, several avenues of self-employment have been opened for the people in the neighbouring areas, specially at railway stations, e.g. transporting and picking up passengers from railway stations to the towns in auto rickshas/three wheelers and tempos. There are 60 stations of Konkan Railway and a number of autorickshas at each station has generated considerable employment in transport sector. This has led to opening of auto workshops at various places along the Konkan Railway route. Local transport, catering and selling refreshment at stations, auto repairing, porter service to passengers for their luggage handling are the other avenues at all the railway stations which have also provided self-employment opportunities to large number of people. Tea stalls, book stalls, provision shops, telephone booths at railway stations were awarded in favour of land losers and local residents. This has also helped in employment generation. The statistics of employees in KR other than casual and the average monthly emoluments per employee are given below:

Table 8.6: Employees in Konkan Railway

Year	*1998-99*	*1999-2000*	*2000-01*	*2001-02*
No. of employees	3382	3382	3414	3435
Average monthly emolument Per employee	8254	9041	9835	9541

Source: Public Enterprises Survey, 1998-99 to 2001-02 published by Department of Public Enterprises, Government of India, New Delhi.

Conclusion

The experience related with resettlement of project affected persons world over reveals that good resettlement prevents impoverishment of affected persons and even reduces their poverty by building sustainable livelihoods. Good resettlement in case of Konkan railway project has been beneficial from economic, social and environmental considerations of the region and has contributed to improved project quality and impact. There was no local resistance to the project due to inadequate resettlement, no political tensions, except for the re alignment controversy in a small section of Goa. With proper planning and implementation, the flow of project benefits was visible right from the construction stage. Our field studies point out that land losers are now better off than they were earlier. This project has brought about "Greater common good" for the region through infrastructure development. Konkan railway project thus, has a favourable impact on the land and people of the Western coastal region of India.

REFERENCES

Barbosa, A, M. (November 2000). Konkan railway on a trying track, *Goa Today*, pp. 12-18.

Carpenter, T.G (1994). *The Environmental Impact of Railways*. Wiley, Chichester.

Cernea, Michael M. Involuntary Resettlement in Development Projects, *World Bank Technical Paper* No. 80.

Handbook on Resettlement: A Guide to good Practice, Asian Development Bank. (1998) Manila.

Involuntary Resettlement: Operational Directive, The World Bank Operational Manual, June 1990.

Karnataka Urban Development and Coastal Environmental Management Project, Report and Recommendations of the President, Asian Development Bank, September, 1999.

Peter Penz *et al.* (April 2001). Economic policy, population displacement and development ethics, Paper presented in a workshop "Towards sustainable development, recent research and findings" organized by Shastri Indo-Canadian Institute, New Delhi.

Pilot project on natural resource accounting in Goa, (2001). (Phase 1) Final Report, Prepared for Directorate of Planning and Statistics, Goa. TERI Project Report No. 99, RD 61, New Delhi.

Resettlement Action Plan, Project Co-ordination Consultancy Service for the Karnataka State Highway Improvement Project, Prepared by Government of Karnataka, November 2000.

Salelkar, S.V. (ed.) *A Treatise on Konkan Railway*, Konkan Railway Corporation Ltd., Navi Mumbai, 1999.

9

Konkan Railway: An Asset for Sustainable Regional Development

A new transport infrastructure stimulates development along its length, and has a long-term impact on socio-economic structure. Konkan railway has upgraded the infrastructure services of the western coast. It has ushered an era of rapid social and economic development in this region. It is difficult to assess the influence the railways have caused towards the progress. But it would be no exaggeration to claim that the Konkan railway has been a very important contributing factor in accelerating the pace of developments, in generating the opportunities for socio-economic changes. It has been observed that projects in transport sector have low returns and long gestation periods. However, contrary to this perception the infrastructure project of Konkan Railway did not have a long gestation period. It was constructed within the record time schedule of about 7 years. The benefits of KR were visible in the coastal region right from the commencement of its construction. CII's National Conference (1997) pointed out that India's transport infrastructure has desperately failed to keep pace with the demands of the growing economy primarily on two accounts—poor infrastructure quality and in-built inefficiency in the system. When examined in the context of Konkan railway, this observation does not hold good. It is true that the one hundred and fifty years old railway system in India awaits for a face lift in all spheres to increase its productivity and to improve customer satisfaction. However, KR is a modern infrastructure of very high quality and standard. Modern technology has been adopted in its various activities. The World Development Report, 1994 points out that "the quality and quantity of the facilities created under infrastructure, in many of the cases do not match with the end users requirements. Poor performance and maintenance of the facilities also results into deterioration and inefficiency in the services". When the performance of Konkan Railway is analyzed in the context of above observations, it is observed that the huge investment made for its construction is paying rich dividends. KR has realized the full scope of its assets through proper use of systems, technology and information technology. The system developed through public financing is effectively serving as a public utility and providing several benefits.

Principles and Guidelines of Sustainable Transport Development and their Application to Konkan Railway

Though the extension of transport system is vital to development, it is apprehended that the ever expanding transportation systems and industrial production are endangering the sustainability of the earth. An environmentally and economically more rational transport system is the need of the hour. The strategy of sustainable development of transport, being a long term solution, provides all inhabitants with better accessibility and brings about much better environmental conditions. Infrastructure services that help the poor also contribute to environmental sustainability. The rail mode helps to meet future transport needs in an environmentally and socially sustainable manner. Any rail development has to be justified as the most appropriate transport solution not only to meet the needs and conditions of our times but also to leave sustainable assets for future generations. Railways offer a sustainable mode of transportation. Konkan railway, a mega infrastructure project is a project of great pride for India and has been praised at international level as an example of sustainable infrastructure development. Konkan railway is a new asset which provides transport services throughout the year in a sustainable manner. Konkan railway is playing a vital role in sustainable development of the Western coastal region. The basic guidelines of integral development of all areas including transport systems were laid down and adopted at the United Nations Conference on Environment and Development—The Earth Summit with the Agenda 21 in June 1992. The principles, guidelines and commitments, adopted by the United Nations member countries at a number of conferences thereafter were gradually put in a concrete form in the field of transport systems development at the UN European conference in Vienna in 1997 with the Vienna Declaration on the Environmentally sustainable transport. The European Forum on Sustainable Transport met in Bad Hofgastein in 1998 and formulated a programme of joint action for the implementation. The main principles, guidelines and objectives of the transport systems development adopted in these documents are applicable to Konkan railway and its role in sustainable transport development of the region. The principles and guidelines of sustainable transport development and their applicability to Konkan Railway is explained below:

(1) The transport system has to be based on a long term vision (30 and more years) of the development of all human activities in interaction with the objectives and strategy of harmonious regional, economic, social and spatial development. Railway improvements are justified as the best long term solution for a particular transport need. Konkan railway construction marks the fulfilment of a long cherished dream of the people on the west coast of India as a lifeline for their economic development. It is based on a long term vision of harmonious development of the coastal region. There is a provision of doubling of this line and increasing the speed up to 160 km per hour as the traffic picks up and the demand increases.

(2) The transport system at the local, regional, national and international level has to provide the inhabitants in these areas with as equivalent as possible access to the place of work, services, supplies, cultural and other activities and to provide the economy with functional links to the market and partners within the region and the rest of the area. Konkan railway provides a missing link of rail transport in the national railway system. It thus brings about substantial improvement in accessibility of the western coast with rest of the country. It not only provides access to places of tourist interest, but links major settlements on the coastal region with rest of the country. From local to regional level, the inhabitants of the western coastal region now have an access to national level. The Konkan railway has opened the markets of Gujarat and Rajasthan to Konkan and Goa and tremendously improved the accessibility of coastal Karnataka and Kerala with rest of India.

(3) By means of traffic nodes and terminals the transport system has to integrate all kinds, forms, technologies of transport into a uniform multi modal transport system with an integral logistic support. It has to encourage environmentally and

economically optimum combinations of transport modes (which are optimal with respect to the national economy and environment). Konkan railway provides such an optimal mode of transport for the national economy. Because of its unique geographical location it encourages environmentally and economically optimum combination of various transport modes—rail, coastal shipping, inland waterways within the region served by it (Fig. 9.1). The coordinated time tables of all the modes of transport can help in applying the 'just in time' principle of delivery and dispatching of goods and serve the passengers travelling by various modes. The stations of Konkan railway are the ideal traffic nodes and terminals for this purpose. They can function as stops for all means of transport enabling easy, quick and safe transfer from one means of transport to another. Distance to railway stations/ to public transport connection, is a major factor in determining their use. The share of public transport rapidly decreases with increasing distance to the station. A high standard of public passenger transport is one of the key conditions for sustainable development of urban and rural settlements which are the principal players in sustainable economic and general development of the country. There is a relatively large use of this public transport by people living near stations. More dwellings and opportunities (jobs) are developing near Konkan railway stations and access time to other stations is reduced because of improved connectivity.

(4) The transport system has to define and implement the measures for a relative reduction of traffic, particularly private car use. Where the modes and scope of transport cause unacceptable pressures on human health and environment, as well as damage to the national economy, the choice of the means of transport should be restricted. The policies of 'transport management' pursued so far, need to be redirected from meeting the demand to restricting the use of transport means. The concept of 'ego-mobility' should be replaced by the concept of 'eco-accessibility'. With the commissioning of Konkan railway there has been a substantial reduction in long distance passenger bus services as more and more train

services are introduced to cope with passenger traffic and ro ro service introduced for goods transportation on Konkan railway route. This has resulted in reduction of vehicular traffic on roads. The withdrawal of vehicles means reduction of accidents and environmental pollution to that extent. The concept of eco-accessibility is realized as the operation of Konkan railway has a beneficial and favourable impact on the environment of the Western coast.

(5) As a part of traffic reduction measures, separate lanes for public transport buses or light railway tracks should be built in the same direction as current heavy private car traffic pattern. The alignment of Konkan railway is almost parallel to National Highway No. 17, the main artery of vehicular traffic on the coastal region and has definitely contributed to reduction of road traffic on this highway. If more dwellings and opportunities (jobs) are located near railway stations, access time is shorter for a number of trips, resulting in a higher share of the train compared to the car. In this way distance/connections to railway stations may result in a higher share of public transport. The stations on Konkan railway are now emerging as focal points of public transport and economic activities. They have witnessed rapid population growth during the decade due to the impetus provided by KR for development activities:

(a) The traditional View

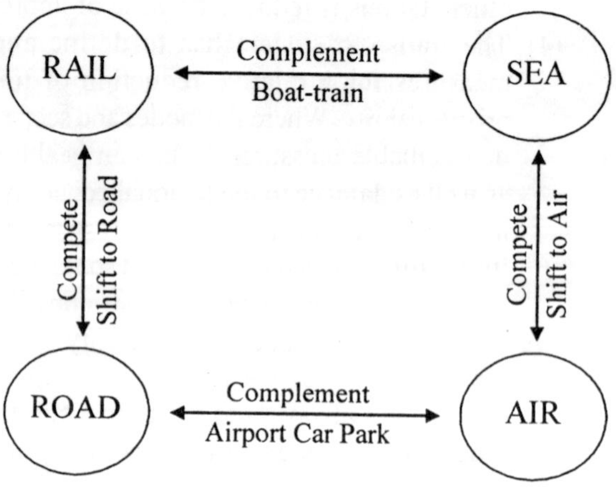

(b) The comtemporary view:

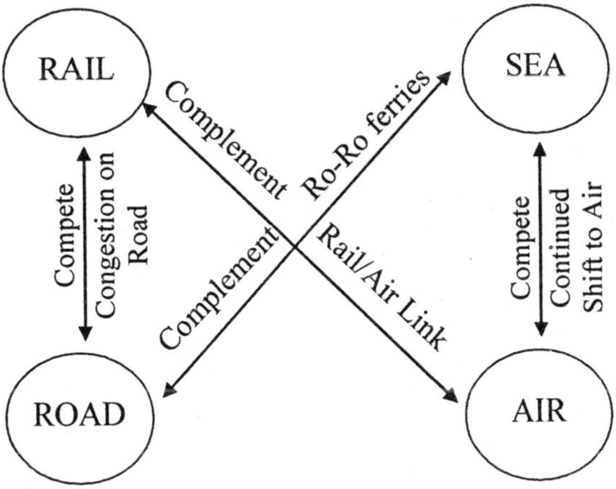

Fig. 9.1 Competition and Complementary in Air, Sea, Road and Air Transport (a) The traditional view; and (b) The Contemporary view.

(6) As a priority, the transport policy, integrated at local, regional and national levels, needs to enforce effective, sensibly priced and physically accessible, comfortable and environmentally friendly public passenger transport with suitable frequency of operation and harmonized time table logistics for bus and rail lines as well as unified tickets for all combinations of transport. Public transport, if in principal, available to all inhabitants is ecologically relatively irreproachable. The operation of passenger services on Konkan railway route fulfil the above criteria of public transport except the unified tickets, which are yet to be introduced. KR has improved access to essential services for the poor. Care is taken to ensure that this infrastructure (KR) protects the interests of the poor by introducing passenger friendly services.

Meeting the Objectives Successfully

With huge investment involved in this railway construction, it seems appropriate to ensure that the system being developed is made as successful as possible. The success of the new railway systems can be measured by observing whether they had attained the objectives that had been set to them. In the study carried out by Mackett and Sutcliffe (2003) the objectives of constructing a new railway system are stated as follows: (i) To improve public transport and to increase public transport usage by reducing car traffic, (ii) To improve land use and urban growth pattern, (iii) To improve access and connectivity, (iv) To stimulate development. The objective related to reducing car traffic includes preventing or relieving traffic congestion and improving the environment by reducing air pollution.

To make a new rail system as successful as possible, certain factors are essential. Physical and socio-economic characteristics of the area influence the success of a system. The area has to be particularly well suited to a rail system. Population density should be medium to high. The towns/cities/stations should be economically viable with location of employment and retail activity mainly at the stations. It is possible to enhance the success of the system by careful planning. This includes providing car parks at stations, restricting car parking in the city centre, locating public development at stations, locating the stations as trip attractors or generators. The railway system has to be integrated into regional plans and bus services have to be integrated with the new system. Operating policies which make the rail system more attractive for passengers and help to increase the patronage are providing frequent train services as per demands of passengers, offering some concessional travel, through effective marketing and advertising and ensuring adequate security on board and at stations. The performance of Konkan railway is evaluated on this background. Our analysis in the preceding chapters reveals that Konkan railway fulfis all the objectives stated above and in the beginning, it has improved the public transport system of the region. It has been successful in attracting passengers and increasing public transport usage. The patronage on new system is increasing. With increased public transport usage, there is reduction in vehicular traffic on roads. This has helped in preventing or relieving traffic congestion and improving the environment of the region by reducing air pollution. It has improved land use and urban growth pattern, improved access and connectivity and stimulated development. Therefore, KR can be termed as a successful railway system. The success of this system is further enhanced by careful planning. The factors essential for the success of a new railway system are also prevalent in this region. Though the area is not very well suited to a rail system, the physical characteristics and climatic conditions of the region justify an all weather transport system. Population density is moderate and not very low. Therefore, more people are benefited. The towns and stations located on KR are emerging as the focal point of economic activity and generating job opportunities. The railway system is integrated into regional plans and bus services are integrated into the new system. Operating policies of KRCL have made the rail system more attractive for passengers and helped to increase the patronage by providing frequent train services as per demands of passengers. The new railway system has helped in improvement in the pattern of urban growth, improvement and development in underdeveloped areas and overall improvement of the environment.

Impacts and Benefits of Konkan Railway

The states in which public sector units (PSUs) are located become the direct beneficiaries and stand to gain manifold in terms of removal of regional imbalances, increased employment opportunities, growth of small scale and ancillary industries, resource mobilization. Konkan railway is a development project in public sector and has benefited the region socially as well as economically. Projects undertaken for development purposes remain meaningless if the local population does not benefit from the development process and continues to get impoverished. KR is rendering vital service to the people on the west coast and providing several benefits. It is very complicated to pinpoint all kinds of direct and indirect effects of a project quantitatively. Konkan Railway has provided the means to exploit the resources of the land and also harness the entire manpower to productive work. The people are exposed to technological changes and hence their living and educational standards are raised.

Strengthening of National Railway Network

Konkan Railway not only gives to the regional transport system its proper balance but also strengthens the national system of railway infrastructure by providing an alternative north south corridor for flexibility in movement. Its diversionary impact helps in relieving the present and future congestion on the Delhi-Madras Grand Trunk route as it provides a north-south link along the coast and minimizes the interchange of journeys. The freight traffic and passenger traffic from north to south which used to move along long roundabout route now take a short cut through KR, thus releasing substantial line capacity on the Indian Railway's congested north-south route. Thus, Konkan railway also brought alternate transport solutions in the country. The line has a great significance from the national defence point of view. The alternative route facilitates manoeuvrability and provide greater resilience to the national system in times of emergency. The line helps to integrate the western coast into a well knit region and to integrate the region with the rest of the country.

Land Use Changes

Care was taken to acquire minimum possible land so that ecology was less affected. Forest land acquired constitutes 0.02 per cent of the total forest cover of our study region and 6.47 per cent of the total land acquired by KRC. Construction of Konkan railway has no adverse impact on the forest cover of this region. The per cent of area under forest increased from 46.84 per cent at the beginning of construction in 1989-90 to 50.42 per cent in 2000-01. More remarkable is the increase in area under dense forest from 13,681 km^2 to 16,900 km^2. According to Forest Survey of India, dense forests now constitute 78.61 per cent of total forests compared to 68.5 per cent a decade back. The gain in forest cover is the outcome of enforcement of stringent rules and regulations for the acquisition of forest land and simultaneous implementation of the various schemes of new plantation and afforestation, under the social forestry and wasteland development projects. Planting tree screens along the rail line was undertaken to reduce pollution. Net area sown has declined in coastal Maharashtra due to urbanization and industrialization.

Konkan Railway: A Better Alternative to Road Travel

In general, railway travel is pleasant, quick and convenient. Looking to the speed and level of comfort for passengers in a train journey, one can safely say that for all long distance journeys especially beyond 500 km and involving more than 10 hours of time rail mode is to be preferred to road. Rail mode has an advantage in all long distance movement. For short distance journeys road is justifiably preferred to rail. Due to excessive rainfall during the monsoon season road transport in this region is not conducive for leisure travel and water transport by steamer from Bombay is discontinued. Konkan railway therefore, provides a better alternative to road travel in the Konkan region and has proved to be extremely popular.

Meeting the Passengers Demands

Infrastructure delivers major benefits in environmental sustainability when it provides services that respond to effective demand and does so efficiently. KR does the same. Transport is a service and must match the needs of the community. The object of any railway system is to move as much traffic as speedily and safely as possible. KR is handling the increased passenger traffic efficiently. The Konkan railway has altered the travel habits of the people on the western coast. By promoting a network of a speedy, safe and cheap transport system, Konkan railway has enabled large masses of people to travel from one part of the country to the other. Customer expectations and competition have grown, especially as alternatives emerged. Due to pressure of competition from roads and other means of transport Konkan railway has to perform at a higher level of efficiency. Konkan Railway is alert and attentive to the demand and needs of trade and industry. In response to the demands of the people of this region KR has introduced new passenger friendly services for short and long distance and number of special trains as per the seasonal and regional requirements from time to time. What makes the line so popular is the cheap fares coupled with a quick and comfortable journey. In fact, the Konkan railway's services are being highly appreciated by passengers. Travelling on the line has gone high tech as they have opted for computerization and information technology in a big way right from the design stage.

Reduction of Vehicular Traffic on Roads

An important environmental benefit of a new railway line is the reduction of traffic from crowded roads. This benefit being stressed in Switzerland and Germany is also relevant in the context of Konkan railway. This rail alignment runs practically parallel to the National Highway No. 17 or west coast highway over a large stretch. Increase in vehicular traffic on this highway had resulted in increase in travel time, number of road accidents, and environmental pollution. By freeing of capacity on the saturated National Highway, Konkan railway is helping in improving the environment. Bus usage along the west coast transport corridor has decreased considerably, as a result of the switch over of the passenger traffic to new rail system. Number of buses, both of Maharashtra State Transport and private transport operators, plying between Mumbai and Ratnagiri and Mumbai and Goa and even to destinations on coastal Karnataka have been substantially reduced or even closed down after the introduction of rail services to these places. All the train services are running to full capacity. Konkan railway helps in reducing the pressure off coastal Goa's over utilised roads. Our study reveals that the Konkan Railway system has been successful in attracting passengers and increasing public transport usage. Konkan Railway has helped to reduce traffic congestion and environmental problems. Konkan Railway's ro ro service of goods transportation also has an impact on reduction of goods carriers plying on roads. Due to reduction in bus and truck usage, there is some relief from traffic congestion on highways and improvement in air quality.

Saving of Time and Distance

Konkan railway has an inherent advantage of providing a faster and safer mode of transport in the coastal region. The construction of KR has benefited the travelling public immensely by reducing their journey time. It has reduced travel time within the states through which it runs and also between the places connected by it. There is substantial reduction of time and distance from places located on KR to Mumbai and other places as shown below in the following table:

Table 9.1: Reduction of Time and Distance

Stations	*Distance in km*		*Saving in Distance*	*Travel time in hrs*		*Saving In hrs*
	Earlier	*New*		*Earlier*	*New*	
Mangalore-Mumbai	2041	914	1127	41	15	26
Mangalore-Ahmedabad	2653	1358	1295	58	32	26
Mangalore-Delhi	3033	2249	784	65	39	26
Cochin-Mumbai	1849	1336	513	36	24	12
Madgaon-Mumbai	629	444	175	20	10	10

Source: *Konkan Railway—A dream comes true*, Press handout by Konkan Railway Corporation Ltd., Belapur, 1998.

The above savings in time are based on a maximum speed of 100 km per hour. Because of this saving, several long distance trains have started on this route. Konkan Railway line has drastically reduced the distance from Delhi and Mumbai to all destinations like Mangalore, Cochin, Trivandrum, etc. The train from Kurla to Ernakulum Netravati Express reaches within 27 hrs by this route whereas Rajdhani Express from Trivandrum to Hazrat Nizamuddin takes only 30 hrs. New Delhi is a mere 24 hours away from Goa if travelling by Rajdhani. Matsyagandha Express from Kurla to Mangalore reaches in 19 hrs. Earlier route via Pune was of long distance and time consuming. Distances between stations in Gujarat and locations in coastal Karnataka and north Kerala have been substantially cut down.

Savings in Fuel Cost

Konkan Railway construction cuts short a circuitous route and brings in real saving to the economy in the form of a reduced generalized cost of transport. Konkan Railway not only saves time and distance for transport users in the large catchments area of the project, but also saves the cost of passengers and goods transportation. Because of the reduction in haulage distance and switch over of road traffic to rail along this route, the savings in fuel cost to the nation was estimated to be around Rs. 200 crore per annum.

A New Asset for Tourism Development

Konkan Railway is ensuring the environmental improvement of this region by properly channelising the tourist traffic to tourist destinations through its passenger friendly services. Konkan Railway has proved to be a new asset for tourism development. The trains running on KR attract a large number of foreign tourists coming from New Delhi and Mumbai to Goa. Counters of pre-paid taxies are opened at Madgaon and other stations to facilitate the onward movement of passengers. A new prestigious luxury train "Deccan Odyssi" has been introduced on this route for foreign tourists. From the number of trains regularly plying between Mumbai and Goa it is evident that Konkan railway is efficiently handling the tourist traffic on this route since its commissioning.

Improved Connectivity of Major Ports

This new railway line provides an effective link with four major ports of Mumbai, JNPT, Mormugao (Goa) and Mangalore on the West Coast with rest of the country. It gives relief to the ports of Mumbai and Cochin as rail link has been provided at intermediate ports like Karwar and Ratnagiri. With infrastructure development, traffic handled by these ports and the share of cargo handled by them to India's total has considerably increased.

Attracting Investment and Promoting Industries

Konkan railway has boosted industrialization, trade and commerce. With the commencement of construction of Konkan Railway several other infrastructure projects were also undertaken on the western coast. A total of 86 major projects are completed in the region after the commissioning of Konkan railway. It is remarkable that the largest number of projects are in infrastructure sector including power projects. Raigad district has emerged as a hub of chemical and pharmaceutical industries. Konkan Railway is assisting in the regeneration of the interior areas, hitherto inaccessible, by attracting investment, not only from the public and private sector, but also foreign direct investment (FDI). The western coast attracted 35 out of 77 FDI proposals in travel and tourism sector approved in India during 1998-2000, i.e. after the commissioning of Konkan railway. Raigad district emerged as a hub of manufacturing activities and Goa attracted investment in consumer goods and tourism industries. FDI is attracted for setting up natural resource based industries, consumer goods and also in service sector. The impact of physical infrastructure and public investment on both private investment behaviour and regional economic development has been found to be highly significant and positive. Studies in China have shown that investments in underdeveloped areas yield much better returns within a short time than investment in the more congested route. This observation holds good for Konkan region as well.

Benefiting the Rural Areas

In most countries, rural areas receive less infrastructure services compared to urban areas. Countries that have made special efforts to provide infrastructure in rural areas have succeeded in reducing poverty dramatically. Konkan Railway passes through rural areas of the coastal region and has made its impact over thereby its very presence. This infrastructure service helps the poor and also contribute to environmental sustainability. Konkan Railway has improved access to essential services for the poor. Konkan Railway has broken down the barriers of isolation surrounding small communities. Care is taken to ensure that this infrastructure (KR) protects the interests of the poor.

Social Benefits

There is need to evaluate social benefits accrued with the completion of projects. The Standing Committee on Railways (1994) observed that Railways should not fail to discharge their social obligations. By providing speedy compensation and employment to land losers, and by undertaking several welfare measures, KR is discharging its social obligations as well. The people are exposed to technological changes and hence their living and educational standards are raised. Konkan Railway has enlarged the facilities for educational development.

Increased Job Opportunities

Transport plays an important role for sustaining economic activities and creating employment opportunities. The experience all over the world reveals that better transport facilities have resulted in increased job opportunities. The growth of non-farm rural employment is linked closely to infrastructure provision. In India it has been observed that in the states where PSUs are located, there has been substantial employment generation, both direct and indirect. The workforce recruited locally in these enterprises constitute a substantial portion of the total employment. Konkan Railway has brought within the reach of common man better opportunities of diversified employment. Konkan Railway has played an important role in expansion of employment opportunities in the backward western coastal region. Konkan Railway generated job opportunities both short term as well as long term for the local people and the residents of the neighbouring areas in the construction and other operational activities. Konkan Railway has expanded market and created effective demand for more and more products so that there is continuing incentive for new work. It has been observed that there is an increase in non-agricultural employment in rural areas after the construction of KR.

Environmental Conservation

Konkan Railway is not posing any serious threat to the fragile eco system of the region. The construction was planned so as to fully preserve the ecology of the area and enhance the beauty of the environment. Extreme care was taken to avoid heritage, religious and community structures during the construction. Konkan Railway has undertaken massive tree plantation programme along the railway line, in the premises of railway stations and staff colonies. An afforestation programme was taken up on embankments and acquired lands. As recommended by the MOEF, the most effective species to absorb the various elements of traffic related pollutants were planted. In addition, a variety of ornamental and fruit bearing trees and teakwood were planted. Use of wooden sleepers is totally avoided and concrete sleepers are used.

The Konkan railway has managed to evolve itself into an economic and cost effective catalyst for growth and prosperity in the Konkan region. To conclude, Konkan railway project has a favourable impact on the land and people of the western coastal region of India. The results of this study clearly indicates that the construction of Konkan Railway has generated positive impact in the region and the same is visible in a very short period of time from its implementation and commissioning. The dream project of the local people is now a reality. The ambitions and hopes are right on tracks of fulfilment. With the resulting benefits which are enjoyed by the common people, the Konkan Railway can be termed as a successful railway system and an asset for regional development.

REFERENCES

Anthony Simoes, (2002). Infrastructure projects, in *Fish, curry and rice, a source book on Goa.* Goa Foundation Publication.

Asian conference on emerging role of the state in the transport sector in the perspective of economic liberalization, (March 1995). New Delhi, Asian Institute of Transport Development, New Delhi.

Emerging issues in transport sector, Conference Papers, (14-16 September, 1989). Organisers: The Chartered Institute of Transport, India, New Delhi.

Guzelj, Tomaz and Ales Sarec, (1998). The concept of a sustainable transport system in the coastal region of Slovenia, *Public Enterprise,* Vol. 16, Nos. 3-4, pp. 329-344.

International Congress on Express Highways Development in India, (December 1998). Background Papers, Organised by Ministry of Surface Transport, Government of India, New Delhi, pp. 14-17.

Mackett, Roger and Ela Babalik, (2003). New urban rail systems: a policy based technique to make them more successful. *Journal of Transport Geography*, 11, pp. 151-164.

Master Plan for Tourism Development in Goa, (1990). Government of Goa, Panaji.

Pilot project on natural resource accounting in Goa (Phase 1) (2001). Final Report, Prepared for Directorate of Planning and Statistics, Goa, TERI Project Report No. 99, New Delhi.

Ranade, P.S., Konkan railway: a new asset for tourism development and environmental improvement in Western India, *Abstract Volume, 29th International Geographical Congress*, Seoul, August, 2000, pp. 436 438.

The State of the Environment, 1980, UNEP, Nairobi, 1981.

Whitelegg, John, (1992). *Transport for a Sustainable Future: The Case for Europe*, Belhaven Press, London.

Alston, L.L. *Railways and Energy*, (March 1984). World Bank Staff Working Papers No. 634, The World Bank, Washington D.C.

India Transport Sector—Long term issues, (March 1995). World Bank Infrastructure Operations Division, Country Department II, India, South Asian Regional Office.

Rao, I.C. (November 1999). Coastal Transportation in India, *Journal of Indian Ocean Studies*, Vol. 7, No.1, pp. 19-37.

Ranade, P.S., (January 2000). Infrastructure development and foreign direct investment on the Western coast of India, *Foreign Trade Review,* Vol.35, No.4, pp. 39-54.

Barbosa, Alexandre, (November 2000). Konkan Railway on a trying track, *Goa Today*, pp. 12-18.

Golias, J.C. (June 2002). Analysis of traffic corridor impacts from the introduction of the new Athens Metro System, *Journal of Transport Geography,* Vol. 10, No. 2, pp. 91-97.

Bibliography

Aitken, Bill (1994). *Exploring Indian Railways*, Oxford University Press, Delhi.

Alston, L.L. (March 1984). *Railways and Energy*, World Bank Staff Working Papers No. 634, The World Bank, Washington D.C.

Androde, Carol (ed.) *The Economic Times Ports of India*, (1999). Bennett Coleman & Co., Mumbai.

Asian Transport Journal, (June 1998) pp. 47-55.

Banerjee Guha, S. (1998). Foreign collaboration in Indian manufacturing during early years of planning (1951-1970): A spatio-temporal analysis, *Geographical Review of India*, Vol. 60(4), pp. 457-479.

Barbosa, A. M. (November 2000). Konkan railway on a trying track, *Goa Today*, pp. 12-18.

Bhat, D.M. (October 2001). Forest dynamics in tropical rain forests of Uttar Kannada district in Western Ghats, *Current Science*, Vol. 79, No. 7.

Bhatia, A.K. (1982). *Tourism development—principles and practice*, Sterling Publishers, New Delhi.

Bilsborrow, R.E. & Pamela DeLargy, (1991). Land use, migration and natural resource deterioration: The experience of Guatemala and the Sudan, in Kingsley Davis,(ed) *Resources, Environment and Population*, Present Knowledge, Future Options, Oxford University Press, New York.

Butler, R.W. (1985). Evolution of Tourism in the Scottish Highlands, *Annals of Tourism Research,* Vol. 12, pp. 371-91.

Carpenter, T.G. (1994). *The environmental impact of railways*. Wiley, Chichester,

Cernea, Michael M. *Involuntary Resettlement in Development Projects*, World Bank Technical Paper No. 80.

Chakravarti, A.K. (1982). *Railways for Developing Countries,* Chetna Publications, New Delhi.

Ellis, Roystone (1989). *India by Rail*, Brand Publication, London.

Ghosh, Arun, Whatever is happening to infrastructure development in India?

Ghosh, B. and Pravir De (1998). Role of infrastructure in regional development: A study over the Plan period, *Economic and Political Weekly*, Vol. 33, Nos. 47 and 48.

Golias, J.C. (June 2002). Analysis of traffic corridor impacts from the introduction of the new Athens Metro system, *Journal of Transport Geography*, Vol. 10, No. 2, pp. 91-97.

Gore, Christopher, (1994). Developing environmental policy on British railways, *Journal of Transport Geography*, 2(3), pp. 204-207.

Goyal, L.C. (Feb.1998). An introspection in some aspects of infrastructure, *Yojana*.

Guzelj, Tomaz and Ales Sarec, The concept of a sustainable transport system in the coastal region of Slovenia, *Public Enterprise*, 1998, Vol. 16, Nos. 3-4, pp. 329-344.

Ha, Myung Shin (2003). A comparison of service quality at major container ports, implications for Korean ports, *Journal of Transport Geography*, 11, pp. 131-137.

Heggie, Ian. *Designing environmentally sound transport projects*, Infrastructure.

Hugo Priemus and Wil Zonneveld (2003). What are corridors and what are the issues? Introduction to special issue: the governance of corridors, *Journal of Transport Geography*, 11, pp. 167-177.

Impact of infrastructure development on ports of India's Western coast, *Foreign Trade Review,* Vol. 37, No. 3 & 4, (October 2002—March 2003), pp. 69-84.

Iqbal, B.A. (1998). *F.D.I. flows to India*, Ajanta Publications, New Delhi.

Jha, C.S., Dutt and Bawa, (25 July 2000). Deforestation and Land use changes in Western Ghats, India, *Current Science*, Vol.79, No.2, pp. 231-237.

Joseph, Anto (ed.) (2002). The *Economic Times Ports of India*, Bennet Coleman & Co, Mumbai.

Krishnaveni, B. and Tapati Mukhopadhyay, Environmental degradation in newly developed urban space in Navi Mumbai—A case study, *The Deccan Geographer*, Vol. 39, No. 1. January to June 2001, pp. 91-100.

Mackett, Roger and Sutcliffe, Ela Babalik (2003). New urban rail systems: a policy based technique to make them more successful, *Journal of Transport Geography*, 11, pp. 151-164.

Modak, S.K. & Nachane, D.M. (June 1980). *West coast transport network, An inter-modal mix*. A Study sponsored by Govt. of Maharashtra, Department of Economics, University of Bombay.

Murphy, A.B. (1992). Western Investment in East-Central Europe: emerging patterns and implications for state stability, *Professional Geographer*, 44(3), pp. 249-259.

Notes, Infrastructure and Urban Development Department, *World Bank*, (1990). pp. 21-25.

Peter Penz *et al*. (April 2001). Economic policy, population displacement and development ethics Paper presented in a workshop "Towards sustainable development, recent research and findings" organised by Shastri Indo-Canadian Institute, New Delhi.

Ranade, P.S. (2000). *Industries and Environment, a study of impact assessment*, APH Publishing Corporation, New Delhi.

Ranade, P.S. (1996). Liberalisation and emerging trends of industrialization in Maharashtra, *Yojana*, 40(3), pp. 16-18.

Ranade, P.S., Infrastructure development and foreign direct investment on the Western coast of India, *Foreign Trade Review*, Vol.35, No. 4, January-March 2000, pp. 39-54.

Ranade, P.S., Konkan railway: a new asset for tourism development and environmental improvement in Western India, *Abstract Volume, 29th International Geographical Congress*, Seoul, August 2000, pp. 436-438.

Rao, I.C. Coastal transportation in India, *Journal of Indian Ocean Studies*, Vol. 7, No.1, November 1999. pp. 19-37.

Rao, V.K.M. Corridor transport networks—an integrated development approach to coastal shipping, *Journal of Indian Ocean Studies*, Vol. 11, No 1, April 2003, pp. 35-48.

Salelkar, S.V. (ed.) (1999). *A Treatise on Konkan Railway*, Konkan Railway Corporation Ltd., Navi Mumbai.

Shivdasani, Menka and Raju Kane (1998). *Konkan railway, a dream come true.* Konkan Railway Corporation Ltd., Navi Mumbai.

Show, Gareth and A.M. Wiliams (1994). *Critical issues in tourism, A geographicla perspective*, Blackwell, Oxford, UK.

Singh Tejveer, *et at.* (ed) (1992). *Tourism environment: nature, culture, economy*, Inter India Publication, New Delhi.

Singh, Biri and R. Chand, Spatial pattern of urbanization in Himachal Pradesh, 1981-91, *Geographical Review of India*, Vol. 62, No. 1, March 2000, pp. 58-70.

Sjoberg, O. and M. Josefson (August 1996). The geography of foreign direct investment in Central and Eastern Europe: a model, Presented in 28th International Geographical Congress, The Hague.

Sreedharan, E. Konkan railway: new frontiers in funding, technology and management. *Asian Transport Journal,* June 1998, pp. 83-91.

Sreedharan, E. (April 1997). Success story of Konkan Railway, *Indian Railways,* pp. 67-72.

Thomas, Jibby (2000). Foreign direct investment in India and Pakistan: a comparative study, Ph.D. Thesis, Jawaharlal Nehru University, New Delhi.

Vaidya, B.C. (ed.) (2003). *Geography of Transport Development in India*, Concept Publishing Co., New Delhi.

Venkitaramanan, S., Infrastructure: Asia's disturbed sector, *The Sunday Times of India*, April 5, 1998.

Wee, Bert van (2002). Land use and transport: research and policy challenges, *Journal of Transport Geography*, 10, pp. 259-271.

Whitelegg, John (1992). *Transport for a sustainable future: the case for Europe*, Belhaven Press, London.

Government Publications

Annual Report, (1998-99). Ministry of Tourism, Government of India, New Delhi.

Basic ports statistics of India (1999-2000). Traffic handled at Indian ports, Transport Research Wing, Ministry of Surface Transport, Government of India, New Delhi.

Census of India, (1981). *Regional Divisions of India: a cartographic analysis*, Series 1 Vols of Maharashtra, Goa and Karnataka.

Census of India, (1991). *District Census Handbooks* of Raigad, Ratnagiri, Sindhudurg (Maharashtra), Goa, Uttar Kannada and Dakshin Kannada district (Karnataka).

Census of India, (2001). Paper 2 of 2001, *Rural Urban Distribution of Population*, Volumes of Maharashtra, Goa and Karnataka states, Office of the Registrar General of India, New Delhi.

Changing profile of India's West coast, Konkan Railway, Publicity hand out of KRCL.

CII's National Conference on *Transport infrastructure emerging scenario*, December 1997, New Delhi.

Developing Maharashtra's Wastelands (1993). Directorate General of Information and Public Relations, Government of Maharashtra, Mumbai, June.

Discover Karnataka's coastal circuit, Brochure issued by Karnataka Tourism, Government of Karnataka, Bangalore.

Economic Division, New Delhi.

Economic Survey, 1997-98, (1998-99). Government of India, Ministry of Finance,

Economic Survey, (2002-2003). Government of India, Ministry of Finance and Company Affairs, New Delhi.

Emerging issues in transport sector, Conference Papers, 14-16 September 1989. Organisers: The Chartered Institute of Transport, India, New Delhi.

Environmental Assessment Summary Report, (December 2000). Project Coordinating Consultancy Services for the Karnataka State Highway Improvement Project, Government of Karnataka.

Environmental Impact Assessment for the section of Konkan railway line in Goa State, Vol. I, Report, March 1992, RITES, New Delhi.

Environmental Impact Assessment Studies for West Coast (Konkan) Railway line Corridor Survey using satellite (IRS-LISS—II) data, Report submitted to RITES. National Remote Sensing Agency, Hyderabad (year not mentioned).

Environmental Master Plan Study for Dakshina Kannada District, (May 1994). Department of Forest, Ecology and Environment, Karnataka State and Danida, Denmark.

Foreign Investment policy of Government of India, October 1995.

Foreign Trade Statistics of India, Principal commodities and countries, (March 1999 and March 1992). Director General of Commercial Intelligence and Statistics, Ministry of Commerce, Government of India, Kolkata.

Gazetteer of India Maharashtra State Kolaba District, (1984). Revised edition, Mumbai.

Google News Alert Konkan Railway.

Handbook on Resettlement: A Guide to good Practice, (1998). Asian Development Bank. Manila.

India by Rails, (1996). Department of Tourism Government of India.

India Infrastructure Report 2001, (2000). Issues in Regulated Market Structure. 3i Network, Oxford University Press, New Delhi.

India Infrastructure Report 2002, (2002). Governance Issues for Commercialization, 31 Network, Oxford University Press, New Delhi.

India Port Report: Ten years of reforms and challenges ahead, (September 2003). I-maritime Consultancy Pvt. Ltd, Navi Mumbai.

India Tourism Statistics, (1998). Ministry of Tourism Govt. of India, New Delhi.

India Transport Sector—Long-term issues, (March 1995). World Bank Infrastructure Operations Division, Country Department II India, South Asian Regional Office.

India, Policies and issues in forest sector development, (June 24, 1993). Report No. 10965—In, Document of the World Bank, South Asia Regional Office.

Indian Railways—Annual Report and Accounts, 1996-97, Ministry of Railways, Government of India, New Delhi.

India—Reducing Poverty, Accelerating Development, (2000). A World Bank Country Report, Oxford.

Infrastructure in India, (November 1996). Economic Intelligence Service, Centre for Monitoring Indian Economy, Mumbai.

Infrastructure, (February 2003). Economic Intelligence Service, Centre for Monitoring Indian Economy, CMIE, Mumbai.

International Congress on Express Highways Development in India, (December 1998). Background Papers, Organised by Ministry of Surface Transport, Government of India, New Delhi, pp. 14-17.

Involuntary Resettlement: Operational Directive, (June 1990). The World Bank Operational Manual.

Karnataka Urban Development and Coastal Environmental Management Project, (September 1999). Report and Recommendations of the President, Asian Development Bank.

Konkan Railway Corporation, Navi Mumbai, Maharashtra Tourism Development Corporation, Mumbai, Wildlife and Ecotourism, Forest Department, Government of Goa, Panaji.

Maharashtra, Key Data, 1999-2000, (1990). The Blue Book, Sumangal Press, Mumbai.

Master Plan for Tourism Development in Goa, Government of Goa, Panaji.

Monthly News Letters of Indian Investment Centre, New Delhi, Monthly issues from 25 April 1998 to 25 April 2000.

Monthly review of investment projects, January to December 2002 issues, CAPEX, CMIE, Mumbai.

PIB Ministry-wise releases Ministry of Railways.

Pilot project on natural resource accounting in Goa, (Phase 1) (2001). Final Report. Prepared for Directorate of Planning and Statistics, Goa. TERI Project Report No. 99, RD 61, New Delhi.

Proceedings of Asian Conference on Emerging Role of the State in the Transport Sector in the Perspective of Economic Liberalization, (March 1995). New Delhi, Asian Institute of Transport Development, New Delhi.

Publicity Pamphlets Issued by Karnataka Tourism, (1966). Government of Karnataka, Bangalore.

Regional Transport Survey of Maharashtra State, Vol. 1. Part 1, Report, Regional Transport Survey Unit, Government of Maharashtra, Bombay.

Resettlement Action Plan, (November 2000). Project Co-ordination Consultancy Service for the Karnataka State Highway Improvement Project, Prepared by Government of Karnataka.

The India Infrastructure Report, (1996). Policy imperatives for growth and welfare. Expert group on the commercialization of infrastructure projects. Published on behalf of Ministry of Finance, Government of India, by Dr. Rakesh Mohan, NCAER, New Delhi.

The State of the Environment, 1980 UNEP, Nairobi, 1981.

Time bound programme for development of Konkan, (1997-2000). Brochure Published by Directorate of Information and Publicity, Government of Maharashtra, Mumbai.

Tourism Policy of Government of India (1985). Lok Sabha Secretariat, New Delhi.

Tourist Maharashtra: A guide to places of interest in Maharashtra, (1984). Maharashtra Tourism Development Corporation, Mumbai.

Website of Konkan Railway Corporation Ltd. www. konkan railway.com

World Development Report 1994, (1994). Infrastructure for Development, The World Bank, Washington D.C.

Yojana, Vol.42, No. 1. January 1998, Republic Day special.

Index